Envisioning a New Racial Grievance Reporting and Redress System for the United States Military

Focused Analysis on the Department of the Air Force

DWAYNE M. BUTLER, SARAH W. DENTON, CHONG H. GREGORY,
ALBERT M. ESPOSITO, IGNACIO A. LARA, LESLIE ADRIENNE PAYNE,
JEANNETTE TSUEI, MICHAEL J. GAINES

Prepared for the Department of the Air Force
Approved for public release; distribution is unlimited.

 PROJECT AIR FORCE

For more information on this publication, visit **www.rand.org/t/RRA1570-1**.

About RAND

The RAND Corporation is a research organization that develops solutions to public policy challenges to help make communities throughout the world safer and more secure, healthier and more prosperous. RAND is nonprofit, nonpartisan, and committed to the public interest. To learn more about RAND, visit www.rand.org.

Research Integrity

Our mission to help improve policy and decisionmaking through research and analysis is enabled through our core values of quality and objectivity and our unwavering commitment to the highest level of integrity and ethical behavior. To help ensure our research and analysis are rigorous, objective, and nonpartisan, we subject our research publications to a robust and exacting quality-assurance process; avoid both the appearance and reality of financial and other conflicts of interest through staff training, project screening, and a policy of mandatory disclosure; and pursue transparency in our research engagements through our commitment to the open publication of our research findings and recommendations, disclosure of the source of funding of published research, and policies to ensure intellectual independence. For more information, visit www.rand.org/about/research-integrity.

RAND's publications do not necessarily reflect the opinions of its research clients and sponsors.

Published by the RAND Corporation, Santa Monica, Calif.
© 2024 RAND Corporation
 is a registered trademark.

Library of Congress Cataloging-in-Publication Data is available for this publication.

ISBN: 978-1-9774-1262-1

Cover: U.S. Air Force photo by Airman 1st Class Brooklyn Golightly.

About This Report

In May 2020, a viral video of the police killing of an African American civilian, George Floyd, sparked widespread outrage, protests, and civil unrest across the United States and abroad. It also led to many societal efforts to reexamine racial injustice in America, including in the U.S. military. This report informs military efforts to improve diversity, equity, and inclusion (DEI) in the armed forces through an examination of policies and structures that constitute the racial grievance system. Racial and ethnic bias, abuse, prejudice, harassment, discrimination, and injustice threaten the military's ability to meet its mission, and they create vulnerabilities that adversaries can exploit. Negative impacts could span military recruitment, retention, unit cohesion, readiness, performance, leadership, talent management, and the health and welfare of the force. Thus, it is imperative that racial grievances are brought swiftly to the attention of military leaders and that leaders address the issues fairly, effectively, and in a timely manner. This report identifies gaps, ambiguities, inconsistencies, and reported problems in the military racial grievance system and offers recommendations for improvement and further evaluation.

The research reported here was conducted as part of a RAND Project AIR FORCE (PAF) initiative to support DEI within the Department of the Air Force. Oversight of the initiative was provided by Ray Conley, with research concept formulation funding provided by the Department of the Air Force for PAF-wide research. It was conducted within the Workforce, Development, and Health Program of RAND Project AIR FORCE as part of a fiscal year 2021 project, "Envisioning a New Racial Grievance Reporting and Redress System for the U.S. Military."

RAND Project AIR FORCE

RAND Project AIR FORCE (PAF), a division of the RAND Corporation, is the Department of the Air Force's (DAF's) federally funded research and development center for studies and analyses, supporting both the United States Air Force and the United States Space Force. PAF provides the DAF with independent analyses of policy alternatives affecting the development, employment, combat readiness, and support of current and future air, space, and cyber forces. Research is conducted in four programs: Strategy and Doctrine; Force Modernization and Employment; Resource Management; and Workforce, Development, and Health. The research reported here was prepared under contract FA7014-16-D-1000.

Additional information about PAF is available on our website:
http://www.rand.org/paf/

This report documents work originally shared with the DAF on February 12, 2021. The draft report, dated February 2022, was reviewed by formal peer reviewers and DAF subject-matter experts.

Funding

Funding for this research was made possible through the concept formulation provision of the Department of the Air Force–RAND Sponsoring Agreement. PAF uses concept formulation funding to support a variety of activities, including research plan development; direct assistance on short-term, decision-focused Department of the Air Force requests; exploratory research; outreach and communications initiatives; and other efforts undertaken in support of the development, execution, management, and reporting of PAF's approved research agenda.

Acknowledgments

We would like to thank Ray Conley; Anthony Rosello; and our peer reviewers, Clarence Johnson and Elicia John, for their constructive feedback on earlier versions of this document. Their insights and advice helped to improve our report. We also extend our sincere appreciation to Chandra Garber for her communications expertise provided throughout the project.

The insights documented in Appendix A were made possible only by the generosity of the 57 subject-matter experts who voluntarily shared their time, knowledge, and points of view with us. We also appreciate insights offered by Miriam Matthews and Jaime Hastings' assistance with panel notes. Laura Miller assisted with the expert panel interviews and interpretation and provided feedback on earlier drafts of this material.

Summary

Issue

The U.S. military is considered to have been a leader in racial integration and advancement for people of color. However, more than 70 years after President Harry Truman's Executive Order requiring equality of treatment and opportunity in the military regardless of race, color, religion, or national origin (Executive Order 9981, 1948), deep-rooted disparities in such areas as discipline and career opportunities persist between White service members and service members from minority racial and ethnic groups (Stewart, Pell, and Schneyer, 2020; U.S. Department of the Air Force [DAF], Inspector General, 2020). The military's grievance system should play a major role in identifying and addressing instances and sources of bias, prejudice, and discrimination, and military leaders should be well prepared and motivated to effectively target individual, cultural, and systemic issues that contribute to such problems. A better understanding of the weaknesses and strengths of the military's racial grievance reporting and redress system is needed to understand where and how it can be improved to encourage racial grievance reporting, facilitate timely and effective responses, and promote a more inclusive environment to better support the careers, satisfaction, and well-being of minority service members.

Approach

To understand the extent to which racial injustice occurs and hurdles that might complicate resolution, our project team employed a mixed-methods approach that included a search of the research literature, government reports, and statistics; a policy analysis; and eight subject-matter expert (SME) panels and two individual interviews that consisted of current and former military personnel who were working or studying at RAND, and RAND researchers with highly relevant expertise. In late 2020, the SME panels solicited insights on types of racial grievances experienced in the U.S. military, grievance reporting options within and external to the military, factors related to willingness or ability to report racial grievances, and actions that can be taken to resolve the problems. Nearly all of the 57 experts who participated are current or former military personnel. Most participants served as officers in the active component, but personnel with enlisted and reserve component experiences were also included. Some participants were researchers whose expertise stemmed not from direct military service but from extensive research in relevant topics or disciplines, such as racial integration in the military, industrial and organizational psychology, and human resource management. While the research team consulted with DAF personnel—with the U.S. Air Force being the focus—to formulate recommendations with broader U.S. Department of Defense implications, no direct consultation or interviewing occurred with Office of the Under Secretary of Defense for Personnel and Readiness team members as part of our approach. The research team applied a doctrine, organization, training, materiel, leadership and education, personnel, and facilities (DOTMLPF)-based framework analysis to the observations and data from the document review, policy analysis, and SME panels to formulate our recommendations.[1]

Throughout this report, we take a system-of-systems approach: Rather than exploring the reporting and redress systems as singular structures, we discuss racial grievance reporting and redress as an interrelated and independent system of systems. We updated this report in late 2022 to incorporate relevant policy changes that occurred before publishing.

Thematic Gaps and Recommendations

The findings and recommendations discussed in the report are summarized in Table S.1. The report also maps these thematic gaps and recommendations to the military's framework for the domains of DOTMLPF.

Table S.1. Summary of Thematic Gaps and Recommendations to Improve the Racial Grievance Reporting and Redress System in the U.S. Military

Thematic Gaps	Recommendations
1. Historical patterns of disparities are embedded in policy and practice.	1. Address historical disparities in policy and practice.
a. There is a willingness to appreciate problems but less willingness to identify and address root causes.	a. Identify and address the root causes of disparities in discipline.
b. The services' ability to identify, monitor, and evaluate racially motivated misconduct, military equal opportunity (MEO), complaints, and equal employment opportunity (EEO) complaints varies.	b. Standardize reporting data and implement oversight for documentation to ensure quality and consistent data to support trend analysis and reporting of results.
c. Minorities lack trust in the chain of command's willingness or ability to address racial grievances.	c. Promote services to support the aggrieved (e.g., chaplain or counseling) and provide alternative reporting channels.
2. A proficiency gap exists in the commanders' role in the racial grievance reporting and redress system.	2. Incorporate checks and balances on commander decision authority.
a. Commanders are not experts in diversity, equity, and inclusion (DEI).	a. Ensure that commanders have appropriate education and training to address DEI issues.
b. Commanders have much decision authority with limited accountability.	b. Consider an objective, independent body consisting of equal opportunity and DEI SMEs to investigate racial grievance reports and recommend options for redress to commanders.[a]
c. Execution and compliance across commanders and across cases are inconsistent.	c. Document, monitor, and evaluate MEO, EEO, and DEI SME advice and commander actions.

[1] The research team used the DOTMLPF framework to identify gaps and needs in the racial grievance and redress system in order to determine solutions and recommendations to address DEI and MEO and redress system shortfalls to help the DAF achieve the professed goals of a DAF grounded in equality and justice. (See Air Force Instruction [AFI] 10-601, 2013, p. 52; and Chairman of the Joint Chiefs of Staff Instruction 5123.01H, 2021.)

Thematic Gaps	Recommendations
3.1. Cultural barriers present obstacles at all levels.	3.1. Mitigate cultural barriers at all levels.
a. The current application of DAF core values might reinforce hostile culture at the individual and unit levels.	a. Publicly commit to changing institutional culture, and assess individual attitudes and unit cultures.
b. Perceptions of risk to one's professional career might outweigh willingness to report.	b. Improve education on retaliation and hostile work environment to set expectations.
3.2. The policy language is vague.	3.2. Strengthen policy language.
a. Suggestive rather than directive language allows for subjective interpretation and application of the racial grievance reporting and redress process.	a. Set standards for fair, equitable, and nondiscriminatory behavior, and use an accountability mechanism in the event of failure to meet those standards.
b. The DAF's definition of diversity might not address all racially motivated misconduct.	b. Encompass a broader variety of problems, including unit-level issues, and change language to say that violations *will* (rather than *may*) result in disciplinary action.
3.3 The system's organization is complex.	3.3. Reduce organizational complexity.
a. Organizational positioning is at the three-letter support staff level, and no single organization has the necessary investigative authorities.	a. Adopt an organizational framework that is similar to the Secretary of the Air Force, Inspector General, which includes oversight by an independent investigative authority.
b. There is a lack of policy or guidance that holistically describes the racial grievance reporting and redress system.	b. Provide guidance that holistically describes the military racial grievance reporting and redress system, as well as the available reporting and support channels.
c. There is little transparency into the racial grievance reporting system for potential complainants.	c. Increase transparency; notify the General Court-Martial Convening Authority when necessary; and, in egregious cases, share information publicly.

NOTE: The thematic gaps in the left column correspond to the recommendations in the right column.
[a] As of September 30, 2022, the DAF revised its Equal Opportunity Program instruction, specifying that MEO and EEO data collection is to be managed within the DAF Equal Opportunity Information Technology System. Notably, the updates to Department of the Air Force Guidance Memorandum (DAFGM) to the Department of the Air Force Instruction (DAFI) 36-2710 (2021) from previous renditions provides increased insight into use of the data.

Contents

Figures and Tables

Figures

Tables

Chapter 1. Introduction

The U.S. military was one of the first institutions in the country to racially integrate and attempt to promote advancement for people of color. As of 2019, 31 percent (413,764) of U.S. active component personnel report being a racial minority, and 17 percent (221,554) report being of Hispanic or Latino ethnicity (Office of the Deputy Assistant Secretary of Defense for Military Community and Family Policy, 2019, p. iii).[2] In the reserve component, 26 percent (213,572) report being a racial minority, and 14 percent (108,635) report being Hispanic or Latino (Office of the Deputy Assistant Secretary of Defense for Military Community and Family Policy, 2019, p. v). However, more than 70 years after President Harry Truman's Executive Order requiring equality of treatment and opportunity in the military regardless of race, color, religion, or national origin (Executive Order 9981, 1948), deep-rooted disparities in discipline and career opportunities persist between White service members and service members from minority racial and ethnic groups.

In September 2020, the U.S. Department of Defense (DoD) published its most recent instruction on equal opportunity (EO), which states that "the MEO [military equal opportunity] program will ensure that Service members are treated with dignity and respect and are afforded equal opportunity in an environment free from discrimination" and that the program will

> process, resolve, track, and report MEO prohibited discrimination complaints . . . prevent and respond to prohibited discrimination . . . hold leaders at all levels appropriately accountable for fostering a climate of inclusion . . . and prevent retaliation against Service members for filing an MEO discrimination complaint. (DoD Instruction [DoDI] 1350.02, 2020, p. 4)

Although codifying such policy represents an important step forward—shortly followed by the first DoDI published to guide the implementation of diversity and inclusion across the enterprise (DoDI 1020.05, 2020)—it is critical that the department and the services take active measures to create an inclusive command climate. According to the *2017 Workplace and Equal Opportunity Survey of Active Duty Members*, 17.9 percent of active duty members indicated experiencing racial or ethnic harassment and/or discrimination within the previous year, with Black (31.2 percent) and Asian (23.3 percent) members reporting more instances of discrimination than White members (12.7 percent) (Office of People Analytics, 2019).[3] In

---

[2] In accordance with U.S. Office of Management and Budget directives, "Hispanic/Latino" is the only ethnicity tracked and is not considered a minority race designation; thus, it is reported separately from race. Data are rounded to whole numbers.

[3] The cited document was not published until August 2019, two years after the survey was conducted. These Congressionally mandated reports are outlined in 10 U.S.C. § 481, requiring a quadrennial survey to assess racial and ethnic relations.

comparison, the *2015 Workplace and Equal Opportunity Survey of Reserve Component Members* found that 14.8 percent of reserve component members experienced racial or ethnic discrimination in the year prior to the survey (Office of People Analytics, 2018).[4]

Ideally, the military's grievance system should play a major role in addressing racial and ethnic disparities by identifying and remediating instances and sources of bias, prejudice, and discrimination. DoD solicits the following information through Workplace and Equal Opportunity surveys:

- indicators of positive and negative trends for professional and personal relationships among members of all racial and ethnic groups
- the effectiveness of DoD policies designed to improve relationships among all racial and ethnic groups
- the effectiveness of current processes for complaints on and investigations into racial and ethnic discrimination.

Data from the surveys augment other growing evidence of an increasing loss of confidence in the grievance system. The May 2020 police killing of civilian George Floyd led to mass protests and prompted societal reexaminations of racial injustice in the United States, including within the military. In December 2020, the U.S. Department of the Air Force (DAF) Inspector General (IG) completed an Independent Racial Disparity Review (RDR), which found the following two compelling issues related to grievance reporting (DAF IG, 2020):

- Service members expressed a lack of satisfaction regarding MEO, with special emphasis on the process of referring cases back to the chain of command.
- A majority of Black DAF members expressed a lack of trust in their chain of command to address racism, bias, and unequal opportunities.

Respondents in the aforementioned Independent RDR, as well as in other fora, have openly opined that the racial grievance reporting and redress system is inadequate for targeting individual and systemic barriers to creating opportunities for change. To target abuse, harassment, discrimination, misconduct, prejudice, and injustice in the military, as well as to promote diversity, equity, and inclusion (DEI), racial minority grievances must be swiftly brought to the attention of leadership and addressed fairly and effectively. This includes ensuring that the command climate promotes dignity, respect, and inclusion for all racial and ethnic minority service members. Unaddressed issues can have negative impacts on recruitment, retention, readiness, performance, talent management, unit cohesion, military leadership, and the health and welfare of the force. The services have faced recent problems with missed recruiting goals and gaps in mission critical occupations. Moreover, public and congressional attention has focused on racial disparities in the military justice system.

---

[4] The cited document was not published until July 2018, three years after the survey was conducted.

Objective

The objective of this research was to propose a more robust, cohesive, comprehensive, and effective racial grievance reporting and response system that would promote DEI in national security through better identification and targeting of the factors that work against it.

Method and Analytic Approach

To understand the extent to which racial minority grievance reporting occurs, as well as hurdles that might complicate the process, our project team employed a mixed-methods approach, including review of policies and programs, review of government reports and statistics, search of the research literature, and subject-matter expert (SME) panels. The research team used framework analysis as the primary approach, wherein we employed the framework based on doctrine, organization, training, materiel, leadership and education, personnel, and facilities (DOTMLPF) to treat observations and data to formulate our recommendations. The attributes of the framework and our application of it are described below.

Policy and Document Review and Analysis

The research team analyzed DoD and DAF policies, government reports and statistics, and related research to discern the methods and processes designed to address racial grievances. The literature and policy reviews revealed gaps that enable persistent discriminatory practices (whether intentional or not) and opportunity deficits experienced by minority service members in the DAF. The key policy documents that the team surveyed are

- DoDI 1020.05, *DoD Diversity and Inclusion Management Program*, 2020
- DoDI 1350.02, *DoD Military Equal Opportunity Program*, 2020
- DoDI 1020.03, *Harassment Prevention and Response in the Armed Forces*, 2020
- DoD Directive (DoDD) 1020.02E, *Diversity Management and Equal Opportunity in the DoD*, 2018, as amended
- DoDI 1440.1, *The DoD Civilian Equal Employment Opportunity (EEO) Program*, change 3, 1992
- DoDI 1350.03, *Affirmative Action Planning and Assessment Process*, 1988
- Air Force Instruction (AFI) 90-301, *Inspector General Complaints Resolution*, 2020
- Department of the Air Force Guidance Memorandum (DAFGM) to the Department of the Air Force Instruction (DAFI) 36-2710, *Equal Opportunity Program*, 2021
- Air Force Policy Directive 36-27, *Equal Opportunity (EO)*, 2019
- AFI 36-7001, *Diversity and Inclusion*, 2019
- Secretary of the Air Force (SAF), Office of the Inspector General Complaints Resolution Directorate, *Commander Directed Investigation (CDI) Guide*, 2016
- U.S. Air Force Doctrine Volume 2, *Leadership*, 2015.

The historical analysis provided in Chapter 2 allowed the research team to contextualize the findings about the system.

Subject-Matter Expert Panels

This project team also gained insights on the military racial grievance reporting and redress system by convening eight SME panels and two individual interviews that consisted of military personnel who were working or studying at the RAND Corporation, in addition to RAND SMEs with extensive research experience on the nexus of race and military affairs.[5] The panelists were diverse in military service, military rank (although most were officers), military career specialty, race and ethnicity, age, gender, time spent in service, service component, command experience, and more.

We ensured that our military participants were either currently serving or had separated within the past ten years. Our nonmilitary panelists comprised experts on the military's racism and discrimination policies, racial and ethnic harassment and discrimination more generally, workforce diversity and inclusion, organizational culture and climate, social psychology, human resource management, and other relevant specialties.

While the research team consulted with DAF personnel—with the U.S. Air Force being the focus—to formulate recommendations with broader U.S. Department of Defense implications, no direct consultation or interviewing occurred with Office of the Under Secretary of Defense for Personnel and Readiness team members as part of our approach.

For more information on our SME panels, see Appendix A.

DOTMLPF Analytical Framework for Qualitative Analysis

Researchers on this project used the DOTMLPF framework to identify gaps and needs in the racial grievance and redress systems and to promote a DAF grounded in equality and justice. We based the build of our analytic framework on the management areas of the DOTMLPF framework because of its broad applicability in terms of its holistic lenses and its familiarity in DoD contexts.

The following shows how the team applied the DOTMLPF framework in examining the racial grievance and reporting system:

- **Doctrine:** Review the doctrine and policy for key roles, responsibilities, and relationships and identify strengths, gaps, conflicts, and mismatches of processes and language.
- **Organization:** Assess current organizations, organizational positioning, and the ranks of key players.
- **Training:** Assess training effectiveness and opportunities for improving as required (time), and identify barriers to training implementation.
- **Materiel:** Review program resourcing; assess the information technology (IT) system used to report and track racial grievance processes to identify trends and areas of concern before they become a problem; track repeat offenders, commander response, victim

---

[5] U.S. Marine Corps service member, interview with authors, November 19, 2020; and former member of the U.S. Army, interview with authors, December 3, 2020. SME panels were held in November 2020.

redress, and other common report items; and examine assessment tools that can be used to gauge attitudes and unconscious bias and train to EO and greater needs.

- **Leadership and education:** Evaluate commanders' authorities and the bases of those authorities, examine how leaders are held accountable and what checks and balances are in place regarding their management of complaints, and determine whether leadership is providing consistent guidance and execution across all commands and echelons.
- **Personnel:** Examine the role of culture in promoting or detracting from how racial grievances are handled, and assess whether offices are handling complaints and are properly staffed, trained, and empowered.
- **Facilities:** Identify the adequacy of facilities and infrastructure (physical and digital) to support the processes.

The framework allowed the team to assess the following:

- how the management areas of the framework relate to one another
- how each element is expected to be an improvement to or preserve a strength of the current system and the basis for that expectation
- which elements need to be implemented first or could be quick wins versus longer-term efforts
- any additional research that might be needed to address shortfalls or gaps.

Organization of This Report

The remainder of the report begins with a discussion in Chapter 2 on the historical context of the racial grievance and reporting system. Then, in Chapter 3, we discuss commanders' roles and needs with respect to maintaining good order and discipline, as well as the tools they should have to fully support the racial grievance and redress system. In Chapter 4, we move into a discussion of the system itself—both how complainants enter it and navigate the process and the challenges that emerge therein. Finally, in Chapter 5, we summarize thematic gaps and present recommendations from the analysis of all data streams through our application of the DOTMLPF framework. In Appendix A, we share SME insights gleaned from our panels. In Appendix B, we go into more detail about the history of race in the U.S. military. In Appendix C, we offer a comparative analysis of the definitions of diversity in the U.S. military services.

Chapter 2. Historical Context

> If it can be shown that the Negro is given an equal opportunity with the White man to qualify for commissioned grades, and that only his lack of qualifications prevent his commission in the higher grades or in combat units, then social and political demands of the administration can be resisted. (Ely, 1925)

Discussion of the historical context of racial grievance reporting in the military generally focus on one of two primary topics: (1) the circumstances surrounding a need for racial grievance reporting and (2) the process of grievance reporting. An examination of the historical policies and practices that underpin the DAF's racial grievance reporting is required to understand the state of the system and address the perception held by some of the majority demographic group, including those at the highest levels of government, that there are no racial disparities. Conversely, members of minority groups have detailed how racial disparities do exist in the DAF, as demonstrated in more than 27,000 single-spaced pages of free text comments in response to a DAF IG request in 2020 and in more than 16,900 single-spaced pages of free text comments in the RDR released in September 2021 (DAF IG, 2021). Contradictory views on the existence of racial disparities are evidenced not only in these reports but also in federal policies. In the span of less than a year, Executive Order 13985 (2021) repealed Executive Order 13950 (2020), which sought to limit federal contractors from discussing "divisive concepts," such as inequities in race and sex, in diversity and inclusion trainings. Executive Order 13985 recognizes that "entrenched disparities in our laws and public policies, and in our public and private institutions" have prevented EO.

The December 2020 IG Report of Inquiry and RDR formally acknowledges the existence of racial disparities between White airmen and airmen from minority racial and ethnic groups but refrains from identifying the underlying conditions and root causes (DAF IG, 2020). Although the six-month assessment of the 2020 RDR completed root-cause analyses of the majority of identified disparities, there remain cases in which some proposed initiatives lack sufficient root-cause analysis necessary to implement effective change and in which insufficient data has meant an inability to conduct root-cause analysis (DAF IG, 2021). Further investigations and literature exploring *why* these disparities exist and *how* they became embedded in the current racial grievance reporting system are needed. Thus, the research team developed a framework for understanding the historical context through two relevant entry points: disparities in opportunities and disparities in discipline.

In this chapter, we focus on the historical aspects of what are now considered triggers for reporting racial grievances in a current context but were common practices in the past. It is important to consider this bleak past as we address triggers that are perhaps latent manifestations in the current racial environment and the corresponding reporting processes.[6] For a more detailed discussion of the historical context of racial and ethnic disparities in opportunity and discipline in the military, see Appendix B.

Disparities in Opportunities

The Civil War to World War II

An examination of the history related to denied opportunities illuminates how racist policies and practices have shaped the military demographically and structurally and how these patterns manifest in the military today. In this section, we outline relevant points related to denied access to opportunities throughout military history that provide the necessary context to understand the need for racial grievance reporting processes. Of note, the historical trends demonstrate that the denial was institutionalized and systemic in both policy and practice. The following are key findings from this historical policy analysis:

- The policy and practice of denying racial and ethnic minorities from military service began with the fight for U.S. independence in 1775.
- Reversal of policies and practices that denied racial and ethnic minorities the opportunity to serve in the military coincided with manpower shortages during wartime.
- Although these policies formally ceased after World War II, latent bias and discriminatory practices continue.

Prior to President Truman ending segregation in the military after World War II, manpower shortages preempted temporary modifications to policies precluding non-White people from serving in the military at the beginning of every major conflict. During the War of 1812, free and enslaved Black people were allowed to serve as soldiers and sailors; however, when the Treaty of Ghent ended hostilities in 1814, it stipulated that "all possessions" taken during the war "shall be returned without delay," including "any slaves or other private property" (Lousin, 2014, p. 1).

At the onset of the Civil War, mass casualties and a declining number of White volunteers led Congress to pass the Militia Act of 1862, making it legal for "Negro men to enlist in the United States Army for the purpose of constructing entrenchments, or performing camp service or any other labor, or any military or naval service for which they may be found competent" (U.S. Congress, 1862, Ch. 86, Vol. I, p. 494).

By the time the United States entered the war against Germany in April 1917, Black people were barred from the U.S. Marine Corps and the U.S. Army Aviation Corps and limited to such menial positions as messmen, cooks, and coal heavers in the U.S. Navy (Defense Equal Opportunity Management Institute [DEOMI], 2002). U.S. Army policy allowed Black soldiers to serve in any position, but, in practice, the majority were placed in service or supply regiments

---

[6] This chapter and Esposito and Gregory (2021) share some common descriptions of this historical context, as they were written concurrently by the same authors for a similar audience.

serving as stevedores, drivers, engineers, and laborers. During his presidency, Woodrow Wilson began to resegregate parts of the federal government, including the Navy and War Department. However, by April 1917 and the U.S. declaration of war against Germany, President Wilson acknowledged that Black service members were critical to the war effort. The Selective Service Act of 1917 allowed, but did not guarantee, that non-White people would be permitted to serve and allowed the Army to continue its policy of segregating Black units (MacLaury, 2000).

During World War II, military leaders continued to believe that non-White people were unfit for combat or leadership positions and continued to relegate them to segregated labor and service units. The Army upheld a 10 percent quota for Black recruits, and, by 1941, Black soldiers in the Army accounted for 5 percent of the Infantry and less than 2 percent of the Air Corps (Kamarck, 2019, p. 4). In contrast, Black soldiers accounted for 15 percent of the Quartermaster Corps. Similarly, 2 percent of the Navy was Black, and nearly all of them, except for six seamen, were relegated to the Steward Mate Corps; none were officers. However, immense political pressure and mobilization requirements soon generated changes to defense policies, despite considerable resistance from military leaders. In response to public pressure, the Roosevelt administration ordered the War Department to create a Black flying unit. In March 1941, the Air Corps established the all-Black 99th Pursuit Squadron, known as the *Tuskegee Airmen*. However, the Tuskegee Airmen continued to face discrimination. In 1945, base commanders at Freeman Field in Indiana ordered Black officers to sign a statement affirming their acceptance of the segregation of the Officers Clubs. More than 100 Black officers refused, were arrested, and were subsequently released with a reprimand on their records. Of the three Black officers who were court-martialed, one was found guilty and received a dishonorable discharge but was ultimately pardoned in 1995 by President Bill Clinton.

Equal Opportunity Efforts After World War II

> The policy of hiding race-based discriminatory practices under the pretense of offering equal opportunity and merit-based advancement casts a long, dark shadow on the historical experiences of minority service members in the military; its chilling effects are felt to this day. (Esposito and Gregory, 2021, p. 41)

The establishment of EO policy was a response to political pressure for employment of Black soldiers during World War I. A 1925 War Department study titled *Employment of Negro Man Power in War* set the stage for President Truman's 1948 executive order to provide EO within the military, stating that Black soldiers "should be given a fair opportunity to perform the tasks in war for which he is qualified" (Ely, 1925, section IV.5). Executive Order 9981 called on the armed forces to provide "equality of treatment and opportunity for all persons in the armed forces without regard to race, color, religion, or national origin" but did not provide a timeline for desegregation of the U.S. military (MacGregor, 1981).

Executive Order 9981 also established the President's Committee on Equality of Treatment and Opportunity in the Armed Forces, which was chaired by Charles Fahey, and charged the body to examine the potential implications of integration on military efficiencies (Kamarck, 2019, p. 4). The Fahey Committee's 1950 report rejected military leaders' claims that desegregation would impair morale and efficiency in integrated units (Kamarck, 2019). Reluctance among senior leaders to change official policies to desegregate the armed forces was met with manpower demands of the Korean War, which catalyzed racial integration in the services. However, the establishment of Executive Order 9981 was not solely an attempt to rectify the racial discrimination and injustices suffered by non-White minorities in the military. President Truman issued the policy just weeks before launching his reelection campaign, helping him secure the racial minority vote, which typically favored the Republican party, and ultimately the reelection. DoD had announced full integration of the active duty military in 1954, but the following years did not result in a significant reduction of systematic discrimination or unequal treatment (U.S. Department of the Army, 2008).

In 1962, the Gesell Commission found a dissonance between policies and practices. Although armed services policies were not discriminatory as written, practices to improve recruitment, assignment, and promotion to achieve equal treatment of Black service members were lacking. The findings also suggested that base commanders were not taking an aggressive role in identifying and addressing racial discrimination (Kamarck, 2019). Efforts to improve EO resulted in the Civil Rights Act of 1964 and the establishment of the Equal Opportunity Commission to enforce policies that prohibit discrimination based on sex, race, color, national origin, and religion (U.S. Equal Employment Opportunity Commission, 1964).

In 1969, Chief Master Sergeant of the Air Force Thomas N. Barnes introduced Social Actions programs (Halvorsen, 2017). The Social Actions Office contained program elements of drug and alcohol abuse, race relations education, and EO and evolved into what is now known as the DAF's EO Program. Initially a component of the Social Actions Office, race relations education fell to the wayside, and the EO program concentrated on individual grievance reporting and redress (Theus, 1972). Although the military's grievance system should play a major role in addressing racial and ethnic disparities by identifying and remediating instances and sources of bias, prejudice, and discrimination, the Independent RDR conducted by the IG found the following two compelling issues related to grievance reporting:

- service members' lack of satisfaction regarding EO, with special emphasis on the process of referring cases back to the chain of command
- a majority of Black DAF members' lack of trust in their chain of command to address racism, bias, and unequal opportunities.

The six-month update of the RDR also documents the existence of inequitable access to opportunity for racial and ethnic minorities. Nine of 16 disparities that were identified in the December 2020 DAF IG RDR report fall within A1 (manpower, personnel, and services support) policy oversight. Using the DAF's eight-step practical problem-solving model, the AF/A1 performed root-cause analysis and developed action plans and measures of success for each finding. Some proposed initiatives still lack sufficient root-cause analysis or an evaluation of the history of inequity in opportunity to understand why the inequity exists. In other cases, a lack of data prevents the thorough root-cause analysis necessary to establish performance measures (DAF IG, 2021). The AF/A1 team is continuing to iterate on how to address these racial disparities in personnel development and career opportunities. The next RDR aims to better detail how resulting countermeasures have worked to remove racial disparities and ensure EO across service members.

Disparities in Discipline

According to Protect Our Defenders' "Military Justice Overview," the U.S. military justice system was modeled after the British system, in which the commanders were vested with immense power (Protect Our Defenders, undated). Service members subject to military courts-martial were afforded little protection and were often convicted without the assistance of counsel or the protection of a judge. Commanders held authority over the charges and sentencing of the accused, and any conviction appeals had to be made directly to commanders.

The results of this unabated power, coupled with the racist attitudes and beliefs of commanders through World War II, had terrible and disparate impacts on Black service members. Of note are two incidents involving Black soldiers and White townspeople that occurred in Brownsville, Texas, in 1906, and Houston, Texas, in 1917 (Kamarck, 2019). In both instances, evidence showed that townspeople instigated or fabricated an incident leading to riots for which the Black soldiers were held responsible and severely punished.

The Brownsville incident, which involved shots being fired from the direction of the garrison into the town, resulted in dishonorable discharges for 167 Black soldiers; this incident remains the only example in U.S. military history of mass punishment without a trial (Kamarck, 2019). In Houston, the beating of two Black soldiers led to a riot in which 16 White civilians and four Black soldiers died (Jeffrey, 2018). Consequently, 64 Black soldiers from the 3rd Battalion of the 24th Infantry were tried in the largest court-martial and murder trial in U.S. military history. Of those tried, 13 were condemned to death. All 64 soldiers were represented by a single lawyer, and their death sentences were carried out without appeal.

This system came under intense scrutiny during World War II, when more than one million courts-martial were convened, subjecting many U.S. service members to the abusive power of commanders wielding unchecked court-martial authority (Protect Our Defenders, undated). According to military defense lawyer Dwight Sullivan, "during World War II, African Americans accounted for less than 10 percent of the Army. Yet, of the 70 soldiers executed in Europe during the war, 55 (79 percent) were African American" (Death Penalty Information Center, undated). In total, the Army executed more than 140 soldiers, with the majority being Black. These executions were carried out within weeks of a conviction, without the benefit of an independent review of the sentence (Protect Our Defenders, undated).

Moreover, since President Truman's order to end segregation in the military in 1948, racial disparities in discipline have actually increased (Death Penalty Information Center, undated). For example, the military carried out 12 executions from 1954 until the most recent one in 1961; 11 of the 12 executed service members were African American (Death Penalty Information Center, undated). A 2017 Protect Our Defenders report that assessed court-martial and nonjudicial punishment (NJP) data from 2006 to 2015 also found significant racial disparity (Christensen and Tsilker, 2017).

- Black airmen were 1.71 times (71 percent) more likely to face court-martial or NJP than White airmen in an average year.
- In an average year, Black marines were 1.32 times (32 percent) more likely to have a guilty finding at a court-martial or NJP proceeding than White marines.
- Black sailors were 1.40 times (40 percent) more likely than White sailors to be referred to special or general court-martial and 1.37 times (37 percent) more likely to have action taken against them in a case.
- Black soldiers were 1.61 times (61 percent) more likely to face general or special court-martial compared with White soldiers.

The significant differences in punishment between Black and White service members are a key element of the historical context behind the process of reporting racial grievances (see Box 2.1). Despite attempts to create policies to promote EO and integration, the legacy of disparity and bias persists, resulting in ineffective practices toward dismantling inequities.

In this chapter, we outlined the historical context and long-standing patterns in racial policy and practice. In Chapter 3, we discuss the role of commanders in the racial grievance reporting and redress system, given the commander-centric nature of the service. In Chapter 4, we discuss how complainants enter the grievance system and navigate the process and the challenges that emerge therein. In Chapter 5, we discuss this thematic gap in more detail and provide recommendations to address this gap according to our DOTMLPF analysis.

Box 2.1

Finding 1: Historical patterns of disparities became embedded in policy and practice, resulting in latent bias in present-day processes and requirements that can perpetuate racial disparities.

Past policies and practices that lack a deeper, critical analysis and redress of racial disparities, as well as consistent reporting and handling of racial grievances, have set a precedent and tone for current policies and practices. We used the term *latent bias* throughout this report to denote biases that have been embedded in past policies and practices, becoming somewhat hidden over the decades, and that continue to have effects at present.[7] For example, given the DAF's historical policies and practices of racial exclusion, which was followed by racial segregation and discrimination, today's integrated DAF continues to have an overrepresentation of Black officers in support, medical, and acquisition career fields (DAF IG, 2020, p. 88; Esposito and Gregory, 2021, Chapter 4) Thus, latent biases stemming from historical policies and practices, such as commanders' subjective judgments pertaining to performance evaluations, can negatively affect Black officer career progression and promotion opportunities. Moreover, requirements might perpetuate racial disparities if they incorporate commanders' subjective judgments. For example, developmental teams identify and provide special attention to high-potential officers based on current and future requirements.[8]

a. **There has been a historical pattern of documenting racial disparities but stopping short of identifying and addressing root causes.**[9] Numerous studies and reports document racial disparities in the military. The December 2020 DAF IG RDR report detailed various disparities, such as those concerning discipline and promotions. Some very capable organizations with very skilled researchers have investigated why the disparities exist, to no avail. Some reports avoid and deny linking causality to such words as *bias* or *racism*, thereby losing credibility among communities that continue to experience discriminatory policies and practices with evidence to support the existence of the discrimination. Without root-cause analysis, causal factors are not addressed, and racial disparities are not effectively mitigated, allowing them to continue to do harm.

b. **The services vary in their ability to identify, monitor, and evaluate racially motivated misconduct and MEO complaints.** On June 16, 2020, the House Armed Services Committee's Military Personnel Subcommittee held a hearing titled "Racial Disparities in the Military Justice System: How to Fix the Culture" (U.S. House Armed Services Committee, 2020). The key message from that hearing was clear: A lack of reliable, consistent data prevented the services from pinpointing the root causes of these disparities (Robinson and Chen, 2020). Recent claims of a lack of reliable, consistent data on why these disparities exist—given the long history of reports on the topic—suggest a deeper cultural issue tied to latent bias that stems from long-standing denial of opportunity and unfair discipline outcomes in policy and practice. Without a clear expectation to report, collect, and critically analyze racial grievance data, the severity of the disparities throughout history (i.e., both policy-driven bias and latent bias resulting from historical policy changes) persists and results in underserved communities' lack of trust in the command structures and any associated avenues of dissent.

c. **These historical patterns have contributed to lack of trust in the chain of command.** Specifically, the DAF IG 2020 report found that 40 percent of Black service members indicated a lack of trust in their chain of command to address racism, bias, and unequal opportunities (DAF IG, 2020, p. 26). Ineffective actions of two stakeholder groups—commanders and organizations involved in official avenues of dissent—affect this lack of trust and reduce their ability to maintain good order, discipline, and equitable treatment for all. The patterns of the lack of trust were reflected in previous data collection efforts, roundtables for the IG investigation, findings of the RDR, and SME panels conducted in support of this research. Command climate has enormous bearing on whether a service member will ultimately choose to report, with toxic commands eroding faith in the reporting process and redress system.

---

[7] For more information on how we use and understand the term *latent bias*, see Esposito and Gregory (2021).

---

[8] For more information on the role of development teams in officer promotion decisions, see Esposito and Gregory (2021, pp. 52–53).

[9] Note that after this study was completed in September 2021, several root-cause analyses working groups were stood up to identify the root causes of disparities documented in the DAF IG's report on racial disparities.

Chapter 3. The Role of Commanders in the Racial Grievance Reporting and Redress System

> As the commander, it is your responsibility to set the tone, establish priorities, and take the lead. As you meet your daily challenges, remember that ultimately command is not about you or how skilled you are in your Air Force specialty. Command is about accomplishing the mission and taking care of your Airmen. (Air Command and Staff College, 2015, p. vii)

The commander is the linchpin of the military institution.[10] Command is a time-honored role that carries with it the burden of tremendous responsibility. Specifically, commanders must be dually focused on accomplishing the military mission and caring for the people in the organization. Although the mission might come first doctrinally (as demonstrated by this chapter's opening quotation), history is rife with the disastrous consequences of a "mission at all costs" command climate; only by putting one's people first can a commander sustainably and reliably execute mission tasks. That is, command is equally about people and task execution; in informal DAF parlance, this is known as "Mission First, People Always." Nonetheless, as a cross-section of U.S. society at large, the military was not and is not immune to reality's distortions of the moral (or indeed, legal) ideal; not everyone who swore an oath to the nation signed a truly fair pact.

In the previous chapter, we discussed how historical discrimination against Black service members was overt and far-reaching. The policies that enforced or enabled such disparate treatment have improved to a great extent, but more-subtle barriers remain. For example, doctrine and policy surrounding DEI are largely written in suggestive phrasing, using verbs like *should* and *may* instead of *must* and *will*.[11] In addition, the largely unchecked discretion afforded to commanders could allow abstract ideas like *unit cohesion* to cloud or, in some cases, override efforts to promote DEI. In any case, perceived and actual disparities in treatment and opportunity

---

[10] The military is an organization of men and women assembled to serve the nation by managing violence on behalf of the people of that nation. Commanders' responsibilities are executed through leadership and apply to both the mission and the people who accomplish it. Commanders are the linchpin at the center of the military's ecosystem that connect people to mission. For more information, see Air Force Doctrine Publication 1 (2021, p. 4).

[11] The difference between suggestive and directive language is delineated by several federal sources. For example, DoD's *Writing Style Guide and Preferred Usage for DoD Issuances* (2020, p. 4) distinguishes the following verbs:

> (1) Use "must" to denote a mandatory action. (2) Use "will" to denote a required action in the future. (3) Use "may" or "can" to denote an optional action that the actor is authorized to perform (a right, privilege, or power that the actor may exercise at his or her discretion)." In addition, the Federal Aviation Administration's Order JO 7110.65Z (2023, section 2) describes usage as follows: "(a) "Shall" or "must" means a procedure is mandatory. (b) "Shall not" or "must not" means a procedure is prohibited. (c) "Should" means a procedure is recommended. (d) "May" or "need not" means a procedure is optional.

experienced by racial minorities persist to this day. These disparities run counter to instilling a sense of trust in commanders and ensuring that good order and discipline is achieved among all members of the organization—as documented in the RDR (DAF IG, 2020).

We first explore the roles and authorities of commanders as set forth by law, policy, and doctrine. Then we connect commanders' roles and responsibilities to the racial grievance reporting and redress system to set conditions for an exploration of the tools available to commanders. We look at the resources that commanders have to address racial grievances and redress, as well as the means available to commanders for discipline. Commanders are often extremely busy and must simultaneously navigate several complex tasks. They often must make decisions that are not popular with the members of the organization or that involve trading what can or will be done with what will not. The following discussion illuminates commanders' vast responsibilities and the necessity to prioritize actions. Prioritization is often linked to the language of written guidance; clearly and strongly written mandates will be accomplished more often than weakly encouraged suggestions. This makes verb choice extremely important when writing policy to dictate the commander actions. Verbs like *encourage* and *motivate* as opposed to *shall* or *must* make decisions less enforceable.

U.S. Law and Department of the Air Force Doctrine and Policy

The legal authorities and obligations of military commanders are contained under Title 10 of the U.S. Code. 10 U.S.C. § 9233 requires all commanding officers and others in authority in the DAF (emphasis added)

> to show in themselves a good example of virtue, honor, patriotism, and subordination; to be vigilant in *inspecting the conduct of all persons* who are placed under their command; to guard against and suppress all *dissolute and immoral practices*, and to correct, according to the laws and regulations of the Air Force, all persons who are guilty of them; and to take all necessary and proper measures, under the laws, regulations, and customs of the Air Force, to promote and safeguard the morale, the physical well-being, and the general welfare of the officers and enlisted persons under their command or charge.

According to Air Force Doctrine Publication 1, *The Air Force*, there are two fundamental elements of leadership—the mission and the airmen who accomplish it (Air Force Doctrine Publication 1, 2021, p. 4). Although good leaders might not always be in command, all good commanders are effective leaders. Air Force Doctrine Publication 1 defines *leadership* as "the art and science of motivating, influencing, and directing Airmen to understand and accomplish [Joint Force Commander] objectives." Furthermore, "effective leadership transforms human potential into effective performance in the present and prepares capable leaders for the future" (Air Force Doctrine Publication 1, 2021, p. 4). To this aim, as a touchstone of DAF core values, leaders are compelled to ensure that all people are treated with equal respect. Doctrine calls on leaders to ensure an inclusive environment among airmen, to encourage their full participation and development, and to pursue on their behalf opportunities for greater responsibility and growth into higher echelons of leadership.

AFI 1-2, *Commander's Responsibilities* (2014), codifies leadership as an inseparable component of command. Reflecting the doctrine from which it stems, AFI 1-2 outlines the following four commander duties and responsibilities:

- Execute the mission.
- Lead people.
- Manage resources.
- Improve the unit.

Effectively leading people therefore falls under the art of command and requires commanders to "establish and maintain a healthy command climate which fosters good order and discipline, teamwork, cohesion and trust. A healthy climate ensures members are treated with dignity, respect, and inclusion, and does not tolerate harassment, assault, or unlawful discrimination of any kind" (AFI 1-2, 2014, p. 3). Leading people also mandates commanders to conduct quality-of-life engagements with their subordinates to "improve quality of life, promote unit morale, and ensure all members are treated with dignity and respect" (AFI 1-2, 2014, p. 3).

Nonetheless, we must again note reality's departure from this doctrinal ideal; recent U.S. history shows that disparate treatment of minorities—especially racial minorities—persists despite evolving policy. The disparities between races manifest in outcomes in such areas as the severity of disciplinary sanctions and the affordance of career advancement opportunities. By charging both good order and discipline and professional development of subordinates squarely upon the unit commander, the DAF places the chain of command at the crux of myriad institutional programs and processes; racial redress grievance and reporting systems are no exception.

The Commander-Centric Nature of the Racial Grievance Reporting and Redress System

As a commander-centric organization, the military delegates significant authority and responsibility to commanders. In the DAF, this is particularly true for squadron commanders, who are in charge of what General Dave Goldfein, 21st Chief of Staff of the Air Force, described as "the Beating Heart of the Air Force" (Goldfein, 2016, p. 1).

Although squadron command is one of the most desirable positions in the DAF, it is also one of the most demanding. Like all DAF commanders, squadron commanders have the following two overarching functions (Judge Advocate General's School, 2019):

- authority over *people*, including the power to discipline
- responsibility for the *mission* and *resources*.

The Uniform Code of Military Justice (UCMJ) (2019) grants commanders special legal authority to ensure good order and discipline.[12] Such special legal authority allows commanders to "independently impose reductions in rank, monetary fines, and restrictions on subordinates' freedoms to encourage behaviors that support mission success" (Air Command and College Staff, 2015, p. 1).

In turn, the DAF devotes significant resources to ensuring squadron commanders' success. Officers are personally selected for command roles and provided with command-specific leadership education and training. Commanders have expert advisors, support staff, and myriad agency resources to ensure mission execution. Select resources available to commanders include the following (Air Command and Staff College, 2015, pp. 14–21):

- The **Air Force Office of Special Investigations (AFOSI)** provides professional investigative services to all DAF commanders. AFOSI identifies, investigates, and neutralizes criminal, terrorist, and espionage threats to DAF and DoD personnel and resources. AFOSI personnel remain independent of the chain of command to ensure unbiased investigations.
- The **Office of the Area Defense Counsel** provides legal defense services for military members. The certified judge advocate, or military attorney, services as counsel in all actions under the UCMJ, as well as in administrative discharge actions.
- **Chaplains** provide spiritual care and ethical leadership. They are the primary advisor for commanders concerning religious accommodation issues and are the only people on base who offer total confidentiality outside the attorney-client relationship.
- The **Equal Opportunity Office** provides formal and informal complaint processing, counseling, conflict resolution, information referral, and other assistance to military and civilian members "who believe they have experienced sexual harassment or unlawful discrimination based on color, race, religion, sex, or national origin" (Air Command and Staff College, 2015, p. 16). Staff conduct command climate assessments for commanders at all levels.
- The **IG** is responsible to the wing commander for assessing and improving unit operational readiness, nuclear surety programs, and mission support effectiveness of all assigned units. The IG is also responsible for establishing and directing DAF complaints and fraud, waste, and abuse programs.

---

[12] Article 15, UCMJ (codified as 10 U.S.C. § 815) gives commanders the authority to impose NJP. NJP encompasses a variety of possible sanctions, including demotions, fines, additional duties, restrictions, and loss of privileges. In addition, if an offense is deemed contrary to good order and discipline but cannot be tried under specific criminal articles in the UCMJ, it can nonetheless be tried under Article 134, UCMJ: the "General Article." Article 134 (codified as 10 U.S.C. § 934) reads in part:

> Though not specifically mentioned in this chapter, all disorders and neglects to the prejudice of good order and discipline in the armed forces, all conduct of a nature to bring discredit upon the armed forces, and crimes and offenses not capital, of which persons subject to this chapter may be guilty, shall be taken cognizance of by a general, special, or summary court-martial, according to the nature and degree of the offense, and shall be punished at the discretion of that court (UCMJ, 2019, p. A2-48).

In essence, the IG serves as the "eyes and ears" of the wing commander, providing an independent fact-binding body to conduct investigations and serving as an honest broker in complaint resolution. The IG is an investigative body only; it neither determines guilt or innocence nor makes recommendations for punishment or remediation—it exists to provide information to the chain of command for decision making (Air Command and Staff College, 2015, p. 16).

- The **Office of the Staff Judge Advocate (SJA)** provides advice to commanders on military justice and disciplinary matters, as well as civil, contract, and environmental law. A key function of SJA is to manage the wing commander's status of discipline meetings, which all unit commanders and first sergeants attend and which provides an opportunity to gauge disciplinary actions across the wing. "Most SJA officers also host regular squadron commander and first sergeant legal training courses to prepare [commanders] for [their] responsibilities associated with military law" (Air Command and Staff College, 2015, p. 15).
- **Sexual assault response coordinators (SARCs)** are part of a DoD-wide program designed to offer support services to victims of sexual assault. Each base has a SARC to help victims and provide advice and training to base personnel. The SARC accomplishes this mission with the help of victim advocates who are volunteers from different units across the base.

Maintaining good order and discipline requires effectively preventing and addressing individual misconduct and interpersonal conflicts in the unit. However, although commanders have significant resources dedicated to ensuring their success, the breadth and depth of their responsibilities and resources typically do not include expertise on decisionmaking related to effectively addressing and resolving racial grievances. The IG and Equal Employment Opportunity offices—the two offices intimately connected to assisting commanders in addressing individual misconduct and interpersonal conflicts in their unit—do not provide commanders recommended courses of action for addressing racial grievances; they provide only information and support. Furthermore, some issues that commanders might encounter are so sensitive and problematic that additional oversight and/or technical assistance is provided—as is the case with sexual assault. When a sexual assault is reported, the SARC, AFOSI, and counseling resources are rapidly leveraged to address the problem. In addition, the SARC rapidly elevates the incident to command levels above the wing commander. Based on historical and recent trends, racial grievances could be considered a high-risk area requiring similar additional oversight and technical assistance.

Commanders have multiple tools to address individual misconduct within the unit. The vast majority of service member discipline is executed with administrative tools; the DAF calls this *quality force management* because the discipline is considered corrective and rehabilitative rather than punitive. Commanders may, without any legal burden of proof and without the involvement of defense counsel, take action that affects a service member's promotability, freedom of movement, and reenlistment or continued service opportunities. Quality force management can also be used for misconduct that does not expressly violate the UCMJ—for instance, showing up disheveled or failing to maintain fitness standards.

If a UCMJ offense is committed, commanders can choose to impose NJP, commonly referred to by its corresponding UCMJ authority: *Article 15*. NJP essentially allows the commander to quickly restore good order and discipline for minor offenses while affording due process to the accused. The accused member is entitled to Office of the Area Defense Counsel services and can refuse an offer to initiate NJP proceedings by demanding trial by court-martial. Military courts-martial are similar to civilian criminal court proceedings; the burden of proof is the "beyond reasonable doubt" standard. Commanders have three categories of responses in their discipline toolbox (Judge Advocate General's School, 2019):

- **Quality force management** (administrative) actions include

  - letters of counseling, admonishment, and reprimand
  - unfavorable information files (UIFs)
  - control rosters
  - administrative demotion (enlisted only)
  - administrative separation
  - selective reenlistment (enlisted only)
  - performance reports
  - promotion propriety action.

- **Nonjudicial** (Article 15) actions are used solely for disposition of UCMJ offenses. Generally, NJP is reserved for *minor offenses*, meaning those in which the maximum imposable punishment at general court-martial does not include a dishonorable discharge or confinement for greater than one year. DAF members (who are not embarked on a naval vessel) reserve the right to refuse NJP and proceed to trial by court-martial. Nonjudicial actions include

  - restriction to limits
  - correctional custody (enlisted only)
  - extra duties (enlisted only)
  - arrest in quarters (officers only)
  - forfeiture of less than 50 percent pay
  - reprimand
  - reduction in grade (enlisted only).

- **Judicial** (courts-martial) actions consist of summary (minor), special (intermediate), and general (serious) proceedings. Outcomes of criminal proceedings can include

  - imprisonment
  - forfeiture of pay or benefits
  - reduction in grade
  - discharge
  - acquittal.

The commander's discipline toolbox is intentionally designed to accommodate the spectrum of misconduct severity. Unless proceeding via court-martial, commanders are encouraged (but not mandated) to apply the "preponderance" standard of evidence, and members subject to quality force management actions are typically not represented or supported by legal counsel. The benefits of such a system are clear—minor offenses can be deftly resolved while serious offenses can still be justly prosecuted. Nonetheless, the trade-off between speed and thoroughness means that minor offenses have a higher risk of abuse of the commander's largely unchecked discretion (whether intentional or unintentional) (see Box 3.1). Currently, discrimination grievances that lack sufficient legal evidence for courts-martial are listed as items that the commander can, optionally, place in the accused member's UIF (Judge Advocate General's School, 2019).

Box 3.1

Finding 2: The racial grievance reporting and redress process relies heavily on the judgment and training of commanders who have substantial responsibility and broad administrative decision authority. However, reliance on the chain of command can promote inherent biases in resolving racial grievances.

Discrimination on the basis of race is illegal. However, grievances that lack the legal evidentiary burden to proceed via court-martial are often left up to the commander to handle with quality force management (i.e., administrative) tools. Given its emphasis on speed and its design for minor infractions contrary to good order and discipline, quality force management gives commanders wide discretion and, in many cases, ultimate decision authority.

a. **Commanders are not experts in DEI, but they have decision authority and significant responsibility to address racial grievances.** As the organizational nexus, commanders are responsible for all aspects of their unit's culture and climate. Simultaneously, commanders are charged with the execution of their unit's military mission and must also maintain the same technical and professional mastery expected of any airman. Thus, commanders are seldom experts in the decisions they face. Despite the support of a command staff and institutional agencies, commanders exercise near-unilateral authority (and bear almost sole responsibility) on myriad issues spanning the spectrum of operational performance to personal crisis management, including issues surrounding DEI.

b. **There are limited checks and balances on commanders' administrative decision authority with respect to racial grievances.** For example, commanders often have the choice of whether to document an allegation of racial discrimination at all; grievances concerning discrimination and sexual harassment are listed as optional entries in an accused member's UIF. In contrast, a UIF entry is required for any officer who receives a letter of reprimand. A memorandum published on December 21, 2020, includes the interim change to AFI 36-2907, *Adverse Administrative Actions*, which now requires commanders to "track adverse administrative actions such as counseling, admonishment, and reprimand," and collect "demographic data of both members issuing and receiving the adverse administrative action" (Fedrigo, 2020).

c. **Inconsistent execution and compliance destroy faith in command and the racial grievance reporting and redress system.** Discretion introduces inconsistency by design; in theory, it allows commanders to tailor their response to the unique circumstances of each grievance. In practice, it can introduce perceptions of favoritism or apathy, especially because the unit at large seldom has access to all of the information that factored into a commander's decision. These perceptions erode trust and confidence in both the chain of command (indeed, in the commanders themselves) and the larger institutional grievance system.

In this chapter, we outlined the responsibility and authority of commanders to ensure the good order and discipline of their unit, as well as the tools and resources provided to them to prevent and address instances of individual misconduct and interpersonal conflicts. In Chapter 4, we discuss the current state of racial grievance reporting from a systemic perspective and further evaluate the policies and practices that place commanders at the center of the redress process. In Chapter 5, we review the thematic gaps and recommendations based on Finding 2.

Chapter 4. A Closer Look at the Military Equal Opportunity Racial Grievance Reporting and Redress System

In this chapter, we give an overview of key terms and concepts, then explore examples of racial grievances within the context of the system, racial grievance reporting and redress policies, entry points into the system, and navigation through the system.

Overview

Talking about race and racism is far from easy. Nonetheless, it is necessary for defining key terms and concepts to create a structure to ensure that there is a method to process real and perceived racial grievances. We understand that *racial grievance* encompasses complaints of bias, discrimination, harassment, favoritism, or other types of disparity based on race, color, or ethnicity and may be against individuals, groups, policies, processes, climates, or cultures. We also acknowledge the interchangeability of terms. Although *racial grievance* was the common parlance used during the SME panels, one could alternatively use *racial discrimination* or, simply, *racism*.

We understand *reporting* to mean initiating a formal or informal complaint via available channels. *Channels* indicate the means by which the complaint is communicated, such as face to face or through a hotline, and the individuals or organizations to which the complainants report. The *redress system* refers to the initial response to a reported incident, such as an investigation, as well as the resolution, including the steps taken to fix the problem, hold people accountable, and prevent recurrence. *Redress* also refers to the documentation of complaints and headquarters-level oversight and monitoring of these processes.

Examples of Racial Grievances

The military's racial grievance system must be able to differentiate and address a wide variety of issues. During this project, we identified myriad examples of racial grievances. Racial grievances in the military can be categorized into three levels: institutional, unit, and individual. At the institutional level, systemic racial discrimination may occur in biased selection of senior leaders, racially insensitive service policies (e.g., prohibiting traditionally ethnic hair styles), and biased delivery of medical care (e.g., selective issuing of pain medication).

Within the unit, certain inequitable behaviors, policies, and cultures, whether overt or subtle, can result in racial grievances. Superior-to-subordinate transgressions include bias in assignment and opportunities, unequal administration of justice, failure to address individual-level transgressions, hostile command climate, lack of mentoring, and exclusion of minorities from social activities. Racial grievances can involve multiple types of bias (e.g., racialized sexual harassment); may be repeated, persistent actions that perpetuate injustice; and may be easy or difficult to prove (see Box 4.1). Bias can be active and based on favoritism toward one group or hostility against another, or it could be exemplified by a leader failing to act.

Box 4.1

Finding 3.1: Cultural barriers to an effective racial grievance reporting and redress system present obstacles to responding to discriminatory behavior at the individual level and are reflected in latent bias embedded in policy and practice at the unit and institutional levels.

At each organizational level, core values and social or situational norms engender an environment that is not conducive to reporting of racial grievances.

a. **Current application of DAF core values may reinforce a hostile culture, particularly at unit and individual levels.** In a perfect world, DAF policies and processes would rid the ranks of discriminatory members by relying on commanders to ensure the morale, welfare, and health of their units. In practice, however, this structure can reinforce a culture that is hostile to filing a grievance, as the complainant might be perceived to be defying the DAF core value of "service before self." Some may see complaints alleging personal discrimination as a hindrance to the "real" warfighting mission of the service. The intent of the "service before self" core value—placing the needs of the service above personal desires—can be misconstrued as an expectation to sacrifice one's sense of self or dignity for the sake of the institution, and filing a grievance can be perceived as troublemaker behavior. In addition, SMEs described pervasive military norms of toughness and resilience, which can consciously or subconsciously discourage reporting.

b. **Perceptions of risk to professional career might outweigh willingness to report a racial grievance.** People who feel compelled to file a racial grievance are not having their best personal or professional moments. Being at odds with coworkers or one's chain of command can be very stressful and detrimental to career outcomes. The 2020 IG RDR report found that 44 percent of Black service members and 33 percent of White service members who reported a racial grievance to their chain of command experienced some form of reprisal (DAF IG, 2020, p. 104). Low willingness to report stems from a lack of examples of institutional-level resolutions, which prevents the ability to set standards for fair, equitable, and nondiscriminatory behavior and consequences for failing to meet those standards; pervasive retooling of unconscious bias training; revision of policy, rules, regulations, and guidance; and a public commitment to changing institutional culture. For a more detailed discussion of factors that might encourage and discourage complainants to report, see Appendix A.

Racial grievances on the individual level may include bias in writing a performance evaluation, which could have long-term impact on one's career. Peer-to-peer examples include offensive jokes or comments, display of offensive symbols or signs, and racist social media posts or association with online hate groups. Discrimination could be intentional or unintentional actions that may arise from ignorance, could occur on or off duty, and could also involve family members of service members.

Racial Grievance and Redress System Policies

DoD and DAF policies describe prohibited behaviors, responsibilities, and programs to prevent and address racial grievances. In this section, we review policies related to racial grievances and describe how a latent artifact from the historical context described in Chapter 2— the definition of *diversity*—casts a long, dark shadow on the ability to identify such racial grievances as racial slurs and race-based harassment.[13] We also discuss how the use of suggestive and ambiguous language in regulations can contribute to a failure to adequately address and resolve racial grievances (see Finding 3.2). We collected and reviewed the main DoD and DAF policies relevant to the military racial grievance reporting and redress process (e.g., MEO) and searched for program products. Box 4.2 provides racial grievance reporting and redress system policies.

Unlike the DAF, the Army and Navy have no centralized system for tracking MEO complaints. The services have long requested that DoD build a centralized, standardized system that could be updated as policies, reporting requirements, and technologies evolve. DoDI 1020.03 requires that substantiated MEO complaints are annotated on fitness reports or performance evaluations; however, we are unaware of any system to ensure that this is being accomplished so that complaints can inform promotion selection boards and identify repeat offenders (DoDI 1020.03, 2020, pp. 7–8). Anecdotal information suggests that this requirement might not be fully met.

---

[13] As of February 2019, the Air Force defines diversity as "a composite of individual characteristics, experiences, and abilities consistent with the Air Force Core Values and the Air Force Mission. Air Force diversity includes but is not limited to: personal life experiences, geographic and socioeconomic backgrounds, cultural knowledge, educational background, work experience, language abilities, physical abilities, philosophical and spiritual perspectives, age, ethnicity, and gender" (AFI, 2019, p. 3). Furthermore, the DAF tailors its concept of diversity to fit specific circumstances as the law requires, and race, ethnicity, and gender fall under the DAF's conception of demographic diversity. Such varying conceptions of diversity allows the term to be fit-for-function—meaning the DAF is able to utilize the conception that fits a need in a given circumstance. Although this may provide some utility, it might also perpetuate the history of identifying racial disparity issues without addressing them. Moreover, it is not clear in DAF policy whether all conceptions of diversity are addressed or treated equally.

Box 4.2

Racial Grievance and Redress System Policies

DoDD 1020.02E, Diversity Management and Equal Opportunity in the DoD, June 1, 2018, as amended.
Purpose: "[E]stablish policy and assign responsibilities to provide an overarching framework for addressing unlawful discrimination and promoting equal opportunity, diversity, and inclusion through
(1) The DoD Diversity and Inclusion Management Program . . .
(2) The DoD Military Equal Opportunity (MEO) Program . . .
(3) The DoD Civilian Equal Employment Opportunity (EEO) Program . . .
(4) The DoD Civil Rights Program."

The policy aims to maximize the productive capacity of those recruited, hired, retained, and promoted through diversity and inclusion and designates the SAF as the DoD Executive Agent for the Defense Equal Opportunity Management Institute.

DoDI 1020.03, Harassment Prevention and Response in the Armed Forces, February 29, 2020.
Purpose: "In accordance with the authority in DoD Directive (DoDD) 5124.02, this issuance:
- Establishes a comprehensive DoD-wide military harassment prevention and response program.
- Updates military harassment prevention and response policies and programs for Service members.
- Updates harassment prevention and response procedures for Service members to submit harassment complaints, including anonymous complaints; procedures and requirements for responding to, processing, resolving, tracking, and reporting harassment complaints; and training and education requirements and standards.
- Supplements the DoD Retaliation Prevention and Response Strategy (RPRS) Implementation Plan for sexual harassment complaints involving retaliation."

DoDI 1350.02, DoD Military Equal Opportunity Program, change 1, September 4, 2020.
Purpose:
- "Establish policy, assign responsibilities, and provide procedures for the DoD Military Equal Opportunity (MEO) Prevention and Response Program.
- Establish the functions of the Defense Equal Opportunity Management Institute (DEOMI) and the DEOMI Board of Advisors (BOA)" to
(1) "Ensure that Service members are treated with dignity and respect and are afforded equal opportunity in an environment free from prohibited discrimination on the basis of race, color, national origin, religion, sex (including pregnancy), gender identity, or sexual orientation.
(2) Process, resolve, track, and report MEO prohibited discrimination complaints, including anonymous complaints.
(3) Prevent and respond to prohibited discrimination through education and training, reporting procedures, complainant services and support, and appropriate accountability that enhances the safety and well-being of all Service members.
(4) Hold leaders at all levels appropriately accountable for fostering a climate of inclusion that supports diversity and is free from prohibited discrimination.
(5) Prevent retaliation against Service members for filing an MEO prohibited discrimination complaint.
(6) Respond to incidents involving harassment, including sexual harassment."

DoDD 1440.1, DoD Civilian Equal Employment Opportunity (EEO) Program, change 3, April 17, 1992.
Purpose: "Recognize equal opportunity programs, including affirmative action programs, as essential elements of readiness that are vital to the accomplishment of the DoD national security mission." This directive
1.1. "Establish[es] the Civilian Equal Employment Opportunity (EEO) Program, to include affirmative action programs . . .
1.2. Consolidates in a single document provisions of references . . .
1.3. Authorizes . . . establishment of Special Emphasis Programs (SEPs) [Federal Women's Program, Hispanic Employment Program, et al.] . . .
1.4. Establishes the Defense Equal Opportunity Council (DEOC), the Civilian EEO Review Board, and SEP Boards.
1.5. Authorizes the issuance of DoD Instructions and Manuals to implement this Directive."

Box 4.2—Continued

SAF, Office of the Inspector General Complaints Resolution Directorate, Commander Directed Investigation (CDI) Guide, June 1, 2018.
Purpose: "The intent of this guide is to provide commanders and their investigative team members the tools they need to conduct commander directed investigations (CDIs). This guide should not be cited as authority for conducting a CDI, and its use is not mandatory. IOs [investigating officers] should consult with the commander directing the investigation as well as the legal office for specific guidance."
Definition: "The CDI is a tool to gather, analyze and record relevant information about matters of primary interest to those in command. The CDI is an extension of the commander's authority to investigate and to correct problems within the command. Therefore, the CDI is internal to the command concerned."

AFI 36-7001, Diversity and Inclusion, February 19, 2019.
Purpose: "This instruction establishes guidance for diversity and inclusion implementation and management, which enables leaders to leverage diverse organizational talent and an inclusive culture to enhance mission effectiveness."

DAFGM to AFI 36-2710, Equal Opportunity Program, September 2, 2021.
Purpose: "[T]o prohibit and eradicate all forms of unlawful discrimination, harassment, and reprisal, and to foster a positive human relations climate, which promotes the full realization of equality of opportunity to all."

AFI 90-301, Inspector General Complaints Resolution, September 30, 2020.
Purpose: "IGs serve as an extension of their commanders by acting as the commanders' eyes and ears to be alert to issues affecting the organization. . . . IGs . . . execut[e] the complaint resolution process and . . . [train] all members of the organization about IG processes and fraud, waste, and abuse issues. A successful complaint resolution program is designed to enhance the organization's discipline, readiness, and warfighting capability." Types of grievances addressed include reprisals, restrictions, and IG wrongdoing in complaints resolution (the IG at the next-higher level will conduct the complaint analysis); "fraud, waste, abuse or gross mismanagement; a violation of law, policy, procedures, instructions, or regulations; an injustice; abuse of authority; inappropriate conduct, or misconduct; . . . [and] a deficiency or like condition."

Air Force Policy Directive 36-27, Equal Opportunity (EO), March 18, 2019.
Purpose: Implement AF EO policies and in accordance with DODDs. The directive addresses the following grievances:
- *Military:* "race, color, sex (including sexual harassment), national origin, religion, or sexual orientation"
- *Civilian:* "race, sex (including pregnancy, gender identity, and sexual orientation), color, religion, national origin, age, genetic information, disability, or prior Equal Opportunity activity."

Although a series of DoD and DAF policies describe prohibited behaviors, responsibilities, and programs to prevent and address racial grievance, no single document or policy holistically defines and details the racial grievance reporting and redress system. The documentation is complex, ambiguous, and time consuming for a complainant to review. Moreover, unlike sexual assault and harassment, no manual or guidebook exists to lead stakeholders through the process (see Box 4.3). In the next section, we describe the process of reporting a racial grievance and outline barriers to effective management and resolution of complaints.

Box 4.3

Finding 3.2: Policy language is vague with respect to the racial grievance reporting and redress system.

Properly classifying offensive behavior as racial in nature can be difficult. Discriminatory behavior and actions might be multifaceted, and there might be an inability to discern the offender's intent, thereby complicating the reporting process.

a. **Use of suggestive rather than directive language in policy documents governing MEO redress is a key weakness, allowing for subjective interpretation and application of the racial grievance reporting and redress system.** For example, using *may* instead of *will* or *shall* introduces possible variance in commander and staff judgments or interpretations concerning MEO redress complaints. The "zero tolerance" policy for discrimination contains contradictory language, starting with strong language that the policy is "*not* to condone or tolerate unlawful discrimination or harassment of any kind" and to take "immediate and appropriate action . . . to investigate," followed by weak language: "Any Airman, military or civilian, who engages in unlawful discriminatory practices or harassment of any kind *may* face disciplinary action" (emphasis added) (DAFGM to AFI 36-2710, 2021, p. 31). If discrimination based on race, ethnicity, or gender *will not* be tolerated, then disciplinary action must not be optional when individuals are found to have engaged in misconduct. On the contrary, polices for drug use, drunk driving, or sexual assault have more-concrete, directive language for proscribed violations. For example, in reference to harassment, DoDI 1020.03 states, "Commanders and supervisors will . . . take appropriate disciplinary or administrative action when a complaint is substantiated" (DoDI 1020.03, 2020, p. 15).

b. **The DAF's definition of diversity might result in racially motivated misconduct not being addressed.** The DAF broadly defines diversity as a "composite of individual characteristics, experiences, and abilities consistent with the Air Force Core Values and the Air Force Mission" (AFI 36-7001, 2019). Although this definition recognizes diversity in many forms, its expansiveness creates a lack of clarity and a high degree of subjectivity in what is considered discrimination. Is the denial of equality to those with any of these "diverse characteristics, experiences, and abilities" considered discrimination, or can discrimination go unchecked because of this broad, vague definition? What are the forms in which racism and racial discrimination manifest in service members' lives and constitute racial grievances? Civilian organizations have defined racial grievances as including inappropriate racial comments; microaggressions; slurs; jokes; pictures; objects; threats; physical assault; intimidation; institutionally or culturally racist policies, practices, and norms; unequal application of policies based on race; and unequal or biased treatment based on race.[14] These definitions include behaviors, actions, or systems that might not be intended to be racist but harm people of color, as well as intentional racial harassment or discrimination. DoD's Office of Equal Employment Opportunity aims to "[e]nsure that Service members are treated with dignity and respect and are afforded equal opportunity in an environment free from prohibited discrimination on the basis of race, color, national origin, religion, sex (including pregnancy), gender identity, or sexual orientation" (DoDI 1350.02, 2022, p. 4). However, the lack of specificity in the language around *diversity* might mean that a variety of types of misconduct, such as racial slurs, are not considered discriminatory; this can result in unreported racial harassment by service members because policy language does not connect racial harassment to racial discrimination. Thus, by not clearly defining unacceptable forms of racial discrimination or connecting discrimination to racial harassment, it is possible that racially motivated forms of harassment might not be perceived as discrimination and treated as such by the MEO/EEO process. For a comparison of how the branches of the U.S. armed forces define diversity, see Appendix C.

Initiating a Racial Grievance Report

The complexity and multitude of channels for reporting a racial grievance, combined with a lack of guidance and transparency, not only make it difficult for a complainant to navigate the reporting and redress process but can also discourage those experiencing discrimination from coming forward (see Box 4.4). We next explore the heterogeneity of different reporting channels,

---

[14] According to the civilian U.S. Equal Employment Opportunity Commission website, race and color discrimination and harassment can take many forms and be embedded in policy and practice. For more information, see U.S. Equal Employment Opportunity Commission (undated).

misaligned incentive structures, and implications of pursuing different types of reporting within the redress process.

Box 4.4

Finding 3.3: The organization of the DAF's racial grievance reporting and redress system is complex.

Although complexity often occurs because a large number of components of a system is interacting in multiple ways and following local rules, the complexity found in the redress system appears to be due to a lack of higher-level organizational direction and attention. We found complexity in areas of organizational structure, instruction and standard protocol, documentation, and resolution, which obfuscates the reporting and redress system. This gap is compounded by Finding 2 in that, although the organizational structure is complex, no single entity has the necessary investigative authorities for an effective racial grievance reporting and redress system.[15]

a. **Complex system structure and organizational positioning at the three-letter support staff level indicate lack of prioritization of racial grievance reporting and redress.** The MEO program resides within SAF/MR, within A1 in DAF, and within the A1 realm at all levels; therefore, it is relegated to a support designation and not central to strategic missions. Program practitioners are primarily middle to lower management in rank (e.g., base installation offices have either a General Schedule (GS)-12 director or a senior noncommissioned officer, and practitioners within the office can start as an E-5). Different EO offices and practitioners exist at higher headquarters (HHQ); these might not be the same office and person at lower levels. This complexity and lack of prominence within the organizational structure implicitly communicates a deprioritization of racial grievance redress. Moreover, by organizational structure the rank and/or grade of the official dedicated to the EO and EEO is the GS-15 director of SAF/MRQ. Correspondingly, the EO representation in the field at the major command (MAJCOM) level are staff advisors to the local commanders with the minimum grade of GS-13 or E8.[16]

b. **Lack of policy or guidance that holistically describes the racial grievance reporting and redress system makes it difficult for complainants to navigate the process.** The research team found no manual or guidebook to lead stakeholders through the process, such as those that exist for sexual assault and harassment. No defined standard exists on how to hold people accountable. Moreover, no guidance or minimum standards on appropriate levels of disciplinary or administrative action are outlined; these actions are left to commander interpretation. A complainant has little guidance or precedent to reference when needing to report a racial grievance.

c. **For a potential military complainant, there is little transparency into the scale and scope of the system.** Prior to the 2021 implementation of GAFDM to AFI 36-2710, there was a limited feedback loop within the military context for racial grievances, resulting in complainants not always being informed of the ultimate decision of an investigation. Complainants might learn of disciplinary actions via criminal judgments or visible administrative actions, but they are not always privy to this information. In comparison, allegations of sexual harassment require notification up the chain to General Court-Martial Convening Authority (GCMCA) within 72 hours. Installation EO issues are not always communicated to base personnel, contributing to a sense that discrimination and racism are nonissues. Moreover, even if there were more transparency into the scale and scope of the system, the documentation is complex and time consuming to review.

---

[15] Although SAF/IG is responsible for the complaint resolution process and administrative oversight of AFOSI—an independent investigative authority—AFOSI's programs are criminal and counterintelligence, not racial grievances. For more information on AFOSI's investigative programs and capabilities, see Office of Special Investigations (2020).

[16] Updated guidance to DAFGM to AFI 36-2710, 2021, released in July 2023, modified responsibilities and references in the publication associated with AF/A1Q to now apply to the DAF Equal Opportunity Office (DoD, 2023).

Reporting Channels

From the perspective of the aggrieved, potential reporting channels are set forth in policy and correspondingly articulated through formal messaging and informal communications within units. However, visibility is lacking in the processes, actors, decisionmakers, timelines, protections, and possible outcomes. In general, *reporting* relates to providing information about serious wrongdoing that one has experienced or has become aware of at their workplace or place of duty. Relatedly, a *reporting channel* is the communication mechanism used to relay the information onward. Figure 4.1. depicts the ecosystem of racial grievance reporting channels within and outside the military.

Figure 4.1. Racial Grievance Reporting Channel Ecosystem

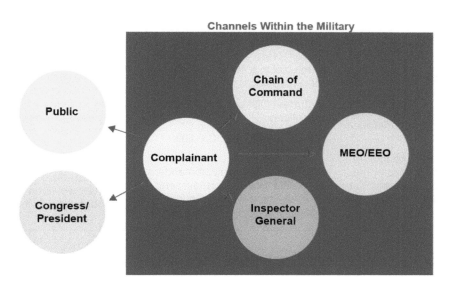

Racial grievance reporting and redress is a *system of systems*—meaning that the grievance reporting and redress system exists within the broader ecosystem of MEO, military justice, and chain of command. A complainant has five channels through which to report their racial grievance—the Office of Equal Employment Opportunity, the IG, and the chain of command channels within the military; the public; and congressional or executive channels that exist outside the military racial grievance reporting process.

The decision to file a report is the first of many steps that the complainant must follow to see a grievance through to its resolution. There are some differences in the systems for military complainants and DAF civilian complainants; however, if the complaint is received by the installation's Office of Equal Employment Opportunity, the general process is similar.

Types of Reporting

There are three ways the MEO/EEO system receives complaints (see Table 4.1). A member can choose to file a complaint informally, formally, or anonymously. The report types differ in the organizational level of the office receiving the complaint, as well as the variety of possible remedies available. Of note, anonymous complaints do not guarantee the protection of the identity of the complainant once the investigation is opened; if the complaint includes enough information to open an investigation, the complainant's identity may be discovered or deduced during the investigative process.

Table 4.1. Military Equal Opportunity Reporting Channel Processes

Informal EO Complaints Process	Formal EO Complaints Process	Anonymous EO Complaints Process
Submitted to: The lowest level appropriate	**Submitted to:** A commander, IG, or staff designated by the military or referral to a MEO professional for processing	**Submitted to:** Air Force Personnel Center/EO, IG, hotline, etc.
Includes: Alternative dispute resolution, intervention, notice to cease, or CDI	**Includes:** CDI conducted by an EO practitioner	If a complaint contains sufficient information (e.g., name, date, and unit) to permit an investigation, the investigation will be initiated by the commander or supervisor in accordance with DoDI and applicable service regulations.
Standard of Proof: Preponderance of evidence	**Standard of Proof:** Credible evidence	
Appeal: If the complaint cannot be resolved within 30 duty days or the complainant is not satisfied with the outcome, the complainant may file a formal complaint.	**Appeal:** Installation commanders, MAJCOM vice commander, and SAF Air Force Review Boards Agency	It is possible that a complainant's identity would be discovered or deduced during the investigation.

The aggrieved may also choose to share racial grievances outside the MEO/EEO channel. For example, allegations of racial abuse can be reported directly to the IG, which can then choose to investigate or refer the case to another agency or the chain of command. Depending on the circumstances and evidence of the case, the military judicial system could become involved. Finally, the aggrieved could choose to bypass the military altogether and air their grievance directly with the media, Congress, or the U.S. public. The complex relationship between the member, their chain of command, and support agencies presents numerous challenges and stresses the grievance process at various points. In the following section, we examine these issues in greater detail.

Navigating the Racial Grievance Reporting and Redress System

Perhaps predictably, the racial grievance process for military members is centered on the chain of command. Members are trained to route essentially any concern to their immediate supervisor and to jump the chain only in extreme cases of impropriety or when the complaint concerns an individual at an intermediate level (e.g., if a complaint concerns the immediate supervisor, the member would report to the section chief). Although there are several institutional offices to support racial grievance redress, their capability to take independent action is severely constrained; often, these external agencies ultimately refer cases back to the member's chain of command for final disposition.

Figure 4.2 depicts the notional life cycle of a racial grievance in the DAF, from the perspective of an individual military complainant (civilian DAF members have additional recourses, such as EEO or suing the DAF).

The first choice the complainant faces is whom to tell. This could include directly reporting to the chain of command (the "textbook" response) or pursuing external agency support. Of these external agencies, MEO is the most tailored for racial grievances; the MEO program exists specifically to "hold leaders at all levels appropriately accountable for fostering a climate of inclusion that supports diversity and is free from prohibited discrimination" (DoDI 1350.02, 2020, p. 4). But military members also have access to agencies designed with broader mandates, including the IG (charged with investigating waste, fraud, abuse, and whistleblower retaliation) and legal recourse through the military justice system (because racial discrimination violates federal and military law). Ultimately, the member could also choose to air a grievance outside any military process whatsoever—complaining to the media, Congress, or the U.S. public at large.

As Figure 4.2 shows, who the member chooses to tell (within the military) might ultimately be of little consequence; the chain of command is inevitably involved. Often, even reports to external agencies are returned to the command for final action. For example, MEO specifically prescribes the results of its investigation into racial discrimination to be disclosed to the command for complaint resolution; referring the case to the chain of command *is* the exit point of the MEO grievance redress process. Similarly, the IG exists strictly in a recommending role. If the member reports a complaint to the IG, the office has the following four potential courses of action:

- Refer the case directly to the member's chain of command.
- Direct ("assist") the case to a more appropriate external agency.
- Elevate ("transfer") the case to a superior IG (the IG for a higher organizational level).
- Investigate the complaint.

If the IG investigates the complaint, it provides the results of the investigation to HHQ (the headquarters at one organizational level above the original complaint) for action; in any case, both the member and the member's chain of command are notified about the IG's disposition decision. Bookending the grievance process with the chain of command is intentional; after all, the commander has ultimate responsibility for the good order and discipline of their unit, so awarding the commander with discretion and latitude appears logical. But the structure presents challenges if members lack trust in their chain of command or, indeed, if their problem is with individuals in the chain of command.

Given that there are substantial challenges when racial grievances make their way through the chain of command, it is important to understand the nature of these challenges and where they emerge. In Figure 4.3, we explore exactly how information flows from the aggrieved to the action-taker—i.e., the commander.

Figure 4.2. Landscape Map of the Military Racial Grievance Reporting and Redress Process

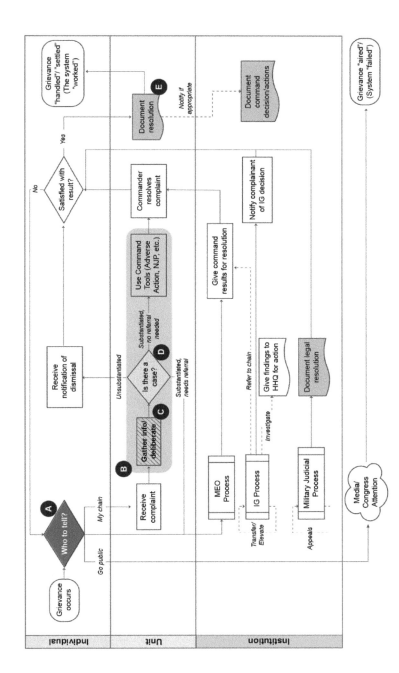

NOTE: (A) At the individual level, the complainant has little clear guidance regarding to whom a particular complaint should be made (purple step). (B) At the unit level, multiple substeps (green box) are involved in the chain-of-command process; every new individual and handoff increase the potential for bias, inattention, loss of context, and other obstacles to a fair resolution. (C) During investigation (hashed step), there is little transparency regarding who in the chain of command has a touch point with the complainant (e.g., will the subject of the complaint be involved? Will those involved in performance evaluations or other career-related decisions be involved?). (D) Many complainants end up in the chain-of-command process (amber step); however, a commander might be perceived to be part of the problem, might not care, might be unwilling to take action, and/or might not have the skills or knowledge necessary to take appropriate action. (E) At the institutional level, not all decisions are documented, and documented (i.e., final) decisions (pink steps) generally lack detail regarding the individuals and deliberations that were part of the investigation and decisionmaking. Without these data, senior leaders cannot effectively understand more-systemic challenges.

33

Figure 4.3. Challenges Within the Chain-of-Command Process

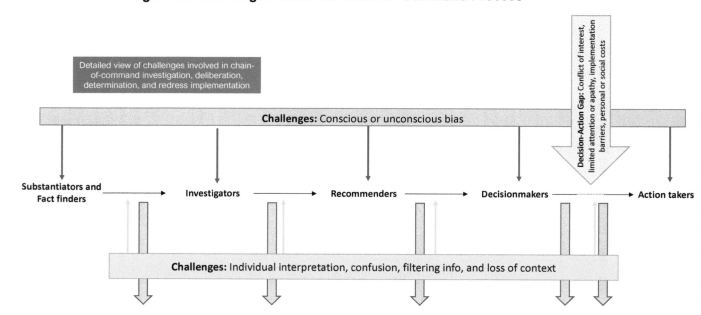

The challenges represented in Figure 4.3 are not unique to the chain of command; they can be generalized to any situation in which information must flow through several individuals before reaching the decisionmaker. In this context, the commander ultimately makes the disciplinary decision with information surrounding the grievance that has been *filtered*, or processed, through various intermediaries. The commander's mental picture of the grievance might not align perfectly with that of the aggrieved member or that of the accused; it is almost certainly missing context that was deemed irrelevant or selectively interpreted at some point along the way. In addition, each individual through which the information is processed introduces their own conscious and unconscious biases, which, in turn, affect what they consider relevant to pass on. Finally, once armed with all of the information available to them, the commander must overcome the *decision-action gap*: the mental barrier between deciding on a course of action and implementing it. This gap often introduces immense pressures, such as strained personal relationships among the commander, the aggrieved, and/or the accused; the institutional incentive structures or "bad optics" associated with the promotability of a commander with an active racial discrimination case; the competing demand for the commander's time relative to other issues; and personal or social costs to implementing a decision that the commander might deem appropriate but that is unpopular with the unit.

In the next chapter, we will discuss the thematic gaps and corresponding recommendations based on the three sets of findings in this chapter. We leverage the DOTMLPF framework to provide a comprehensive, systematic set of recommendations based on each set of findings and their subparts.

Chapter 5. Thematic Gaps, Recommendations, and Conclusions

In this chapter, we consider the findings from the previous three chapters and organize them into concise thematic gaps that map to the recommendations provided later in this chapter (see Table 5.1).

Table 5.1. Mapping of Thematic Gaps to Recommendations

Thematic Gaps	Recommendations
1. Historical patterns of disparities are embedded in policy and practice.	1. Address historical disparities in policy and practice.
a. There is a willingness to appreciate problems but less willingness to identify and address root causes.	a. Identify and address the root causes of disparities in discipline.
b. The services' ability to identify, monitor, and evaluate racially motivated misconduct and MEO/EEO complaints varies.	b. Standardize reporting data and implement oversight for documentation to ensure quality and consistent data to support trend analysis and reporting of results.
c. Minorities lack trust in the chain of command's willingness or ability to address racial grievances.	c. Promote services to support the aggrieved (e.g., chaplain or counseling) and provide alternative reporting channels.
2. A proficiency gap exists in the commanders' role in the racial grievance reporting and redress system.	2. Incorporate checks and balances on commander decision authority.
a. Commanders are not experts in DEI.	a. Ensure that commanders have appropriate education and training to address DEI issues.
b. Commanders have much decision authority with limited accountability.	b. Consider an objective, independent body consisting of EO and DEI SMEs to investigate racial grievance reports and recommend options for redress to commanders.[a]
c. Execution and compliance across commanders and across cases are inconsistent.	c. Document, monitor, and evaluate MEO, EEO, and DEI SME advice and commander actions.
3.1. Cultural barriers present obstacles at all levels.	3.1. Mitigate cultural barriers at all levels.
a. The current application of DAF core values might reinforce hostile culture at the individual and unit levels.	a. Publicly commit to changing institutional culture, and assess individual attitudes and unit cultures.
b. Perceptions of risk to one's professional career might outweigh willingness to report.	b. Improve education on retaliation and hostile work environment to set expectations.
3.2. The policy language is vague.	3.2. Strengthen policy language.
a. Suggestive rather than directive language allows for subjective interpretation and application of the racial grievance reporting and redress process.	a. Set standards for fair, equitable, and nondiscriminatory behavior and use an accountability mechanism in the event of failure to meet those standards.

Thematic Gaps	Recommendations
b. The DAF's definition of diversity might not address all racially motivated misconduct.	b. Encompass a broader variety of problems, including unit-level issues, and change language to say that violations *will* rather than *may* result in disciplinary action.
3.3 The system's organization is complex.	3.3. Reduce organizational complexity.
a. Organizational positioning is at the three-letter support staff level, and no single organization has the necessary investigative authorities.	a. Adopt an organizational framework that is similar to the SAF/IG, which includes oversight by an independent investigative authority.
b. There is a lack of policy or guidance that holistically describes the racial grievance reporting and redress system.	b. Provide guidance that holistically describes the military racial grievance reporting and redress system, as well as the available reporting and support channels.
c. There is little transparency into the racial grievance reporting system for potential complainants.	c. Increase transparency; notify the GCMCA when necessary; and, in egregious cases, share information publicly.

NOTE: The thematic gaps in the left column (gray) correspond to the recommendations in the right column (purple). This color coding flows through the remainder of the DOTMLPF tables in this report.
[a] As of September 30, 2022, the DAF revised its Equal Opportunity Program instruction, specifying that MEO and EEO data collection is to be managed within the DAF Equal Opportunity Information Technology System. Notably, the updates to DAFGM to AFI 36-2710 (2021) from previous renditions provides increased insight into use of the data.

In Table 5.2, we consider how these thematic gaps and recommendations align across the DOTMLPF framework. As the table shows, gaps and recommendations regarding historical patterns of disparities (indicated by gray and purple shading, respectively) touch on all DOTMLPF management areas, with the exception of training. Across all areas, doctrine or policy and leadership are both a gap and solution.

Table 5.2 highlights areas in which the DAF is currently working to improve gaps in its doctrine to address historical disparities in policy and practice, as shown in the lighter shaded areas (under Recommendation 1); improvements to the other gaps have not yet started. Later in this chapter, we discuss each of these thematic gaps and corresponding recommendations individually.

Table 5.2. Summary of Thematic Gaps and Recommendations Across DOTMLPF

Thematic Gap	D	O	T	M	L	P	F	Recommendation
1. Historical patterns of disparities are embedded in policy and practice.								1. Address historical disparities in policy and practice.
2. A proficiency gap exists in the commanders' role in the racial grievance reporting and redress system.								2. Incorporate checks and balances on commander decision authority.
3.1. Cultural barriers present obstacles at all levels.								3.1. Mitigate cultural barriers at all levels.
3.2. The policy language is vague.								3.2. Strengthen policy language.
3.3. The system's organization is complex.								3.3. Reduce organizational complexity.

NOTE: The gray shading (in the top half of each row) indicates those management areas in which the thematic gaps emerge, and the purple shading (in the bottom half of each row) indicates the management area in which the recommendation falls. Lighter shading indicates areas in which the DAF is already taking steps to address these gaps. White cells indicate that we were unable to identify DOTMLPF gaps or associated recommendations. D = doctrine; O = organization; T = training; M = materiel; L = leadership and education; P = personnel; and F = facilities.

DOTMLPF Framework Analysis of Thematic Gaps

DOTMLPF is a variation of the framework used by the Joint Requirements Oversight Council in the Joint Capabilities Integration and Development System to identify nonmateriel solutions to gaps in joint capabilities.[17] Table 5.3 illustrates the elements of each thematic gap (using the numbering from the findings boxes in Chapters 2, 3, and 4) across the DOTMLPF framework.

---

[17] We did not intend to propose any materiel solutions; therefore, DOTMLPF was used as a logical and familiar framework for analysis. For more information about DOTMLPF, see CJCSI 5123.01H (2021).

Table 5.3. Summary of Thematic Gaps Across DOTMLPF Framework

Thematic Gap	D	O	T	M	L	P	F
1. Historical patterns of disparities are embedded in policy and practice.	D	O	T	M	L	P	F
a. There is a willingness to appreciate problems but less willingness to identify and address root causes.							
b. The services' ability to identify, monitor, and evaluate racially motivated misconduct and MEO/EEO complaints varies.							
c. Minorities lack trust in the chain of command's willingness or ability to address racial grievances.							
2. A proficiency gap exists in the commanders' role in the racial grievance reporting and redress system.	D	O	T	M	L	P	F
a. Commanders are not experts in DEI.							
b. Commanders have much decision authority with limited accountability.							
c. Execution and compliance across commanders and across cases are inconsistent.							
3.1. Cultural barriers present obstacles at all levels.	D	O	T	M	L	P	F
a. The current application of DAF core values might reinforce hostile culture at the individual and unit levels.							
b. Perceptions of risk to one's professional career might outweigh willingness to report.							
3.2. The policy language is vague.	D	O	T	M	L	P	F
a. Suggestive rather than directive language allows for subjective interpretation and application of the racial grievance reporting and redress process.							
b. The DAF's definition of diversity might not address all racially motivated misconduct.							
3.3 The system's organization is complex.	D	O	T	M	L	P	F
a. Organizational positioning is at the three-letter support staff level, and no single organization has the necessary investigative authorities.							
b. There is a lack of policy or guidance that holistically describes the racial grievance reporting and redress system.							
c. There is little transparency into the racial grievance reporting system for potential complainants.							

NOTE: The gray shading indicates those management areas in which the thematic gaps emerge, and the lighter gray shading indicates areas in which the DAF is already taking steps to address these gaps.

Next, we discuss the thematic gaps individually and organize them via a DOTMLPF framework. This binning process considers the characteristics of each gap and how they relate to each area in the DOTMLPF framework and illuminates applicable management areas.

In our analysis in Chapter 2, we found that historical patterns of disparities have become embedded in policy and practice. Because of the military's historical policies of racial exclusion, followed by racial segregation and discrimination, coupled with its centuries-old practice of limiting Black service members to unskilled, nonoperational fields, Black officers today are consistently overrepresented in the support, medical, and acquisition fields and are underrepresented in the rated operations fields, which has long-term implications for officer promotions. As of March 2022, there were nearly 340 Black pilots out of roughly 15,000 active-duty pilots in the DAF—that is, about 2 percent, a ratio that has remained consistent for 30 years (DAF IG, 2020; Cohen, 2020).

Racial disparity in military justice and military death penalty cases have also persisted to the present. In 1972, the Secretary of Defense's Task Force on the Administration of Military Justice in the Armed Forces concluded that "the military system does discriminate against its members based on race and ethnic background" and this discrimination is intentional and/or systemic (Office of the Secretary of Defense, 1972, p. 2). Preservice racial and ethnic attitudes were key factors in that "fear, mistrust and suspicion influence the fair administration of military justice and contribute to racial animosity and tension" (Office of the Secretary of Defense, 1972, p. 2). The task force also called out policy and commander efforts, noting an absence of a specific penalty for discriminatory acts or practices and "the gap between the importance the DoD apparently attaches to equal opportunity and human relations programs, and the importance it accords them when it comes to supporting them with money, manpower and good management" (Office of the Secretary of Defense, 1972, p. 51). It further observed that the department's "inconstant support is undermining the credibility of equal opportunity programs among military personnel and adding to existing problems of racial hostilities" (Office of the Secretary of Defense, 1972, p. 51).

In 2017, similar results were found in the Protect Our Defenders *Racial Disparity in Military Justice Report*, which looked at disciplinary data from 2006 to 2015 and found that Black service members across all service branches were substantially more likely to face military justice or disciplinary action than White service members were. In the DAF, Black airmen, on average, were 1.71 times (71 percent) more likely to face court-martial or NJP than White airmen (Christensen and Tsilker, 2017). These results were confirmed by the U.S. Government Accountability Office in 2019 and the DAF IG RDR in 2020 (U.S. Government Accountability Office, 2019; DAF IG, 2020).

As shown in Table 5.4, this finding and its subparts emerge within all management areas except training. Doctrine is implicated by gaps in the willingness to identify and address root causes and in the variation of the ability to monitor and track data related to both the complainants' reports and actions taken during redress.

Table 5.4. Finding 1 DOTMLPF Analysis

1. Historical patterns of disparities are embedded in policy and practice.	D	O	T	M	L	P	F
a. There is a willingness to appreciate problems but less willingness to identify and address root causes.							
b. The services' ability to identify, monitor, and evaluate racially motivated misconduct and MEO/EEO complaints varies.							
c. Minorities lack trust in the chain of command's willingness or ability to address racial grievances.							

NOTE: The gray shading indicates those management areas in which the thematic gaps emerge, and the lighter gray shading indicates areas in which the DAF is already taking steps to address these gaps.

The DAF is already making efforts to address this gap by coordinating teams to conduct root-cause analyses on the disparities identified in the IG RDR report, as detailed in the Six-Month Assessment report released in September 2021 (DAF IG, 2021). This is a critical step in the right direction; however, addressing this gap requires an iterative process with further root-cause analyses, more collection of data, and additional steps to produce desired results, which might include increased financial and human capital resourcing. A related effort is underway at RAND in which researchers are identifying the root causes of racial disparities in military discipline rates among DAF service members. The researchers are examining the extent to which the observed racial disparity in Article 15 proceedings and court-martial rates narrows once detailed controls for service members are accounted for in quantitative analyses. They aim to understand whether this disparity might arise because of discrimination, racial differences in offending rates, or differences in career fields that have higher rates of discipline enforcement.

The ability to identify, monitor, and track racially motivated misconduct will continue to require monetary and personnel resources, as well as appropriate digital infrastructure. Worsening lack of trust in the chain of command emerges as both a personnel and leadership issue. All tables in this chapter show that leadership is a prevailing issue, particularly at the institutional and unit levels.

The DAF should consider the following recommendations to address this gap:

- Continue to identify root causes of patterns of disparities in discipline, including the provision of adequate monetary and personnel resources for the analysis (doctrine, materiel, leadership, personnel).
- Standardize reporting data and implement oversight for documentation to ensure quality and consistent data to support trend analysis and reporting of such results in the future (doctrine, organization, materiel, leadership, facilities).
- Promote services to support the aggrieved—such as the chaplain and counseling—and provide alternative reporting channels outside the chain of command that are not at higher institutional levels like the IG (leadership, personnel).

Table 5.5 shows how the mismatches between commanders' training, authority, and responsibilities relating to DEI emerge across DOTMLPF. Because this finding concerns the commander's role within the racial grievance reporting and redress system, all subparts implicate the leadership and compliance management areas. Doctrine emerges within Findings 2.a. and 2.b. because, although commanders have expert advisors and support staff, they are not required by policy to be experts in DEI, nor does policy provide a mechanism for ensuring that commanders fulfill their responsibility to adequately address racial grievances and guarantee that the complainant is made whole. Organization is also a concern across all subparts; although commanders have significant responsibility and authority for racial grievances, DEI is one of myriad issues they must manage. Moreover, senior commanders should be aware of how junior commanders are responding to racial grievances within their units. In essence, commanders lack the requisite expertise to manage racial grievances, and the EO SMEs who provide guidance to commanders lack the necessary investigative authorities.

Table 5.5. Finding 2 DOTMLPF Analysis

2. A proficiency gap exists in the commanders' role in the racial grievance reporting and redress system.	D	O	T	M	L	P	F
a. Commanders are not experts in DEI.							
b. Commanders have much decision authority with limited accountability.							
c. Execution and compliance across commanders and across cases are inconsistent.							

There are a variety of proposed recommendations the DAF should consider to address this gap:

- Ensure that commanders have the appropriate education and training to address individual DEI issues within their units (doctrine, organization, leadership).
- Consider having objective, independent bodies, including EO and DEI SMEs, investigate racial grievance reports and recommend options for redress to commanders (doctrine, organization, leadership).
- Use and document EO and DEI SME advice, as well as the actions that commanders ultimately take to resolve the grievance (organization, leadership).

In Table 5.6, we show which management areas are implicated in the obstacles presented by cultural barriers to an effective racial grievance reporting and redress system. Leadership and personnel concerns emerge across each subpart at the individual and unit levels. For example, leadership is essential in instilling the DAF's core values within their units according to doctrine and policy, and commanders are best suited to mitigate individual perceptions that reporting a racial grievance might come with risks to complainants' professional careers.

Table 5.6. Finding 3.1 DOTMLPF Analysis

3.1. Cultural barriers present obstacles at all levels.	D	O	T	M	L	P	F
a. The current application of DAF core values might reinforce hostile culture at the individual and unit levels.							
b. Perceptions of risk to one's professional career might outweigh willingness to report.							

To address these concerns, the DAF should consider the following recommendations:

- Senior leadership should publicly commit to changing institutional culture, and commanders should assess individual attitudes that shape their unit culture. Commanders' assessments should supplement, not replace, command climate surveys as a means to convey dedication to creating and maintaining a positive climate within their unit (doctrine, leadership, personnel).
- Commanders should ensure that individuals within their units do not perceive reporting racial grievances as professionally risky; this will require gaining the trust of all personnel within their unit (leadership, personnel).

Of all our findings in this report, Finding 3.2 (shown in Table 5.7) was the most clear-cut after DOTMLPF analyses; both subparts of this finding emerge within the leadership and doctrine management areas. Because each leader (i.e., the individual level) can subjectively interpret the vague language within doctrine and policy (i.e., the institutional level), racially motivated misconduct might not be appropriately addressed at the unit level.

Table 5.7. Finding 3.2 DOTMLPF Analysis

3.2. The policy language is vague.	D	O	T	M	L	P	F
a. Suggestive rather than directive language allows for subjective interpretation and application of the racial grievance reporting and redress process.							
b. The DAF's definition of diversity might not address all racially motivated misconduct.							

The DAF should consider the following to address this gap:

- Set standards for fair, equitable, and nondiscriminatory behavior, as well as an accountability mechanism in the event of failure to meet these standards (doctrine, leadership).
- Change the language within doctrine and policy to ensure that DEI regulations are directive rather than suggestive—that is, violations *will* rather than *may* lead to disciplinary action (doctrine, leadership).

Table 5.8 shows which management areas are implicated by Finding 3.3. Concerns arise within doctrine, policy, and organization across all subparts, and leadership concerns manifest within Finding 3.3.c. The organizational positioning of MEO/EEO professionals at the three-letter support staff level suggests that policy does not dictate appropriate prioritization of the racial grievance reporting and redress system within the DAF. Moreover, there is a lack of policy or guidance that holistically describes the system, which makes it difficult for complainants to both enter the system and navigate their way through the process. This gap is driven by limited transparency into the systems that encompass the DAF's racial grievance reporting and redress system, and transparency is the burden of leadership. Moreover, Finding 3.3 is compounded by Finding 2, because no single entity (e.g., commander or EO professional) has the appropriate combination of subject-matter expertise and disciplinary and redress authority.

Table 5.8. Finding 3.3 DOTMLPF Analysis

3.3. The system's organization is complex.	D	O	T	M	L	P	F
a. Organizational positioning is at the three-letter support staff level, and no single organization has the necessary investigative authorities.		■					
b. There is a lack of policy or guidance that holistically describes the racial grievance reporting and redress system.		■					
c. There is little transparency into the racial grievance reporting system for potential complainants.		■			■		

To address these concerns, the DAF should consider the following:

- Adopt an organizational framework that is similar to the SAF/IG, which includes oversight by an independent investigative authority (doctrine, organization).[18]
- Provide guidance that holistically describes the military racial grievance reporting and redress system (doctrine, organization).
- Increase transparency of the racial grievance and redress process, notify the GCMCA when appropriate, and share information publicly on egregious cases to signal dedication to addressing racial grievances to service members and external stakeholders (e.g., Congress) (doctrine, organization, leadership).

Summary of Recommendations

Lastly, we refine solutions and make specific recommendations. During our SME panels, various recommendations were proposed to mitigate the challenges within the DAF's current racial grievance reporting and redress system. Granted, further research should be conducted to establish the feasibility, achievability, supportability, and timeliness of these proposed recommendations before implementation.

As we mentioned earlier in this report, gaps in the military racial grievance reporting and redress process can manifest within and across individual, unit, and institutional levels. Table 5.9 shows the recommendations to address the thematic gaps identified in this report and across the DOTMLPF framework.

---

[18] Per DAF IG, undated:

> SAF/IG provides administrative guidance and oversight to AFOSI while AFOSI retains its independent authority to conduct its investigative and operational mission. . . . SAF/IG ensures effective inspection and investigative systems, which includes administrative guidance and oversight for the military criminal investigative organization (MCIO) mission conducted by the Air Force Office of Special Investigations (AFOSI).

Specifically, with respect to complaints, SAF/IG is responsible for the complaint resolution program and administrative oversight of its independent investigative programs. For more information on SAF/IG's organizational structure, see DAF IG (undated).

Table 5.9. Summary of Recommendations Across DOTMLPF Framework

Recommendation	D	O	T	M	L	P	F
1. Address historical disparities in policy and practice.	■		■	■	■	■	■
a. Identify and address the root causes of disparities in discipline.							
b. Standardize reporting data and implement oversight for documentation to ensure quality and consistent data to support trend analysis and reporting of results.							
c. Promote services to support the aggrieved (e.g., chaplain or counseling) and provide alternative reporting channels.							
2. Incorporate checks and balances on commander decision authority.	■			■	■		
a. Ensure that commanders have appropriate education and training to address DEI issues.							
b. Consider an objective, independent body consisting of EO and DEI SMEs to investigate racial grievance reports and recommend options for redress to commanders.[a]							
c. Document, monitor, and evaluate MEO, EEO, and DEI SME advice and commander actions.							
3.1. Mitigate cultural barriers at all levels.	■		■		■		
a. Publicly commit to changing institutional culture, and assess individual attitudes and unit cultures.							
b. Improve education on retaliation and hostile work environment to set expectations.							
3.2. Strengthen policy language.	■				■		
a. Set standards for fair, equitable, and nondiscriminatory behavior and use an accountability mechanism in the event of failure to meet those standards.							
b. Encompass a broader variety of problems, including unit-level issues, and change language to say that violations *will* rather than *may* result in disciplinary action.							
3.3. Reduce organizational complexity.	■				■		
a. Adopt an organizational framework that is similar to the SAF/IG, which includes oversight by an independent investigative authority.							
b. Provide guidance that holistically describes the military racial grievance reporting and redress system, as well as the available reporting and support channels.							
c. Increase transparency; notify the GCMCA when necessary; and, in egregious cases, share information publicly.						■	

[a] As of September 30, 2022, the DAF revised its Equal Opportunity Program instruction, specifying that MEO and EEO data collection is to be managed within the DAF Equal Opportunity Information Technology System. Notably, the updates to DAFGM to AFI 36-2710 (2021) from previous renditions provides increased insight into use of the data.

DOTMLPF Framework Analysis of Recommendations

Next, we discuss each recommendation individually in detail.

As shown in Table 5.10, our recommendations to address disparities associated with the historical discrimination and policy discussed in Chapter 2 are threefold. Identifying and addressing the root causes of patterned disparities cuts across the following management areas: doctrine, materiel, leadership, and personnel. Addressing root causes will require leadership buy-in, and identifying these causes will require monetary and personnel resource allocations through policy.[19] The recommendation to standardize racial grievance reporting and redress data and

---

[19] As we mentioned, the DAF is taking steps to identify the root causes within the DAF IG RDR, but it is unclear whether the level of monetary and personnel resources, as well as time, is adequate to address root causes.

implement oversight for documentation to ensure quality and consistent data affects all management areas except training. This recommendation will require monetary and personnel resources to be allocated through policy and requires adequate digital infrastructure and organizational capacity to support longitudinal tracking and monitoring of complaints and resolutions. Furthermore, leaders will be responsible for the standardization of reporting and redress data. Lastly, leaders should promote personnel services, such as talking with chaplains and counselors, to support the aggrieved and provide information on alternative reporting channels to help mitigate eroding trust in the chain of command.

Table 5.10. Recommendation 1 DOTMLPF Analysis

1. Address historical disparities in policy and practice.	D	O	T	M	L	P	F
a. Identify and address the root causes of disparities in discipline.	X	X		X			
b. Standardize reporting data and implement oversight for documentation to ensure quality and consistent data to support trend analysis and reporting of results.	X	X		X	X	X	X
c. Promote services to support the aggrieved (e.g., chaplain or counseling) and provide alternative reporting channels.						X	X

To address Finding 2, identified in Chapter 3 of this report, and the proficiency gaps discussed in Table 5.5, our recommendations emerge within the following management areas: doctrine, organization, training, and leadership (Table 5.11). First, the DAF needs to ensure that commanders have the appropriate education and training to adequately address racial grievances, which will require policy and organizational adjustments. The DAF should also consider establishing an objective, independent body composed of EO and DEI SMEs to investigate racial grievance reports and recommend options for redress to commanders. Recommendation 2.b. requires new policy to establish such a racial grievance investigative body—not dissimilar to AFOSI, which investigates criminal reports and manages counterintelligence programs—within the organization, thereby alleviating some of the pressure on commanders with regard to their significant responsibility for the racial grievance and redress system. Furthermore, the DAF should document the EO and DEI SME advice, as well as whether and how commanders used their recommendations, which will require leadership buy-in and organizational adjustments.

Table 5.11. Recommendation 2 DOTMLPF Analysis

2. Incorporate checks and balances on commander decision authority.	D	O	T	M	L	P	F
a. Ensure that commanders have appropriate education and training to address DEI issues.	X	X	X		X		
b. Consider an objective, independent body consisting of EO and DEI SMEs to investigate racial grievance reports and recommend options for redress to commanders.		X			X		
c. Document, monitor, and evaluate MEO, EEO, and DEI SME advice and commander actions.		X			X		

To address the gaps identified in Finding 3.1, it will be necessary to mitigate cultural barriers across all levels of the DAF. Specifically, the DAF should consider publicly committing to changing institutional culture and assessing individual attitudes and unit cultures. This recommendation cross-cuts the following management areas: doctrine, leadership, and personnel (Table 5.12). The DAF should also consider improving education on retaliation and hostile work environments to set expectations for commanders and the individuals within their units. This will require doctrine and policy to establish the necessary training for leaders and their airmen.

Table 5.12. Recommendation 3.1 DOTMLPF Analysis

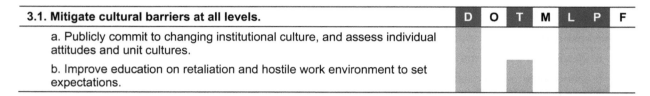

3.1. Mitigate cultural barriers at all levels.	D	O	T	M	L	P	F
a. Publicly commit to changing institutional culture, and assess individual attitudes and unit cultures.	■				■	■	
b. Improve education on retaliation and hostile work environment to set expectations.	■		■		■	■	

The recommendations to address Finding 3.2 are also fairly clear-cut across the DOTMLPF framework (Table 5.13). Doctrine, policy, and leadership must be engaged in setting standards for fair, equitable, and nondiscriminatory behavior, including an accountability mechanism in the event of failure to meet those standards. The DAF should update doctrine and policy to encompass a broader variety of problems, including unit-level issues, and make policy language more directive—that is, violations *will* rather than *may* result in disciplinary action. Leaders will be responsible for conveying this new doctrine and policy to the airmen and in their units, which may help increase trust in the chain of command.

Table 5.13. Recommendation 3.2 DOTMLPF Analysis

3.2. Strengthen policy language.	D	O	T	M	L	P	F
a. Set standards for fair, equitable, and nondiscriminatory behavior and use an accountability mechanism in the event of failure to meet those standards.	■				■	■	
b. Encompass a broader variety of problems, including unit-level issues, and change language to say that violations *will* rather than *may* result in disciplinary action.	■				■	■	

As Table 5.14 shows, our recommendations for Finding 3.3 emerge within the following management areas: doctrine, organization, and leadership, and all three subparts of our recommendation fall within the first two management areas. The DAF should consider adopting an organizational framework that is similar to the SAF/IG, which includes oversight of an investigative authority. As we mentioned, SAF/IG conducts administrative oversight of AFOSI, which is an independent criminal and counterintelligence investigative body that is external to the chain of command.

Table 5.14. Recommendation 3.3 DOTMLPF Analysis

3.3. Reduce organizational complexity.	D	O	T	M	L	P	F
a. Adopt an organizational framework that is similar to the SAF/IG, which includes oversight by an independent investigative authority.	■						
b. Provide guidance that holistically describes the military racial grievance reporting and redress system, as well as the available reporting and support channels.	■						
c. Increase transparency; notify the GCMCA when necessary; and, in egregious cases, share information publicly.	■				■		

The DAF could benefit from adopting a similar structure for racial grievances, which would necessitate new doctrine and policy. The DAF should also consider providing guidance that holistically describes the racial grievance reporting and redress system, detailing all available reporting channels. This guidance would help facilitate our final recommendation: The DAF should increase transparency. Leaders must notify the GCMCA when necessary and should consider sharing information publicly for egregious violations. Such public disclosure of egregious offenses is already commonplace in AFOSI, and the DAF could realize similar benefits in addressing DEI concerns.[20]

---

[20] For more information on the information AFOSI shared on egregious cases during 2020, see Office of Special Investigations, 2020, pp. 11, 13, 15, 19, 21, 23, 25, 29.

Appendix A. Insights from Subject-Matter Expert Panels

Our SME panels were one of several data sources on racial grievance reporting and redress. Although this research is internal to RAND and sponsored by Project AIR FORCE, we emphasized to participants the usefulness of drawing from experiences across the services and asked them to think broadly about these very important issues.

We requested participation from current or recent active and reserve component military personnel who were working or studying at RAND, in addition to RAND SMEs with extensive research experience on the military's racism and discrimination policies, racial and ethnic harassment and discrimination more generally, workforce diversity and inclusion, organizational culture and climate, social psychology, human resource management, and other relevant specialties. Ultimately, 57 of the 62 of the experts to whom we reached out accepted our invitation to participate in the panels. Over a three-week period, our research team elicited perspectives from these 57 participants on different aspects on racial grievance reporting and redress in the U.S. military (see Figure A.1). We began with what was arguably the most difficult question of the set: What might trigger a racial grievance in the military? This was prefaced with a clarification that one need not have personally been the recipient of racist or discriminatory behavior to comment. We next asked about internal (i.e., military) and external communication channels for reporting racial grievances. Thereafter, we asked participants to consider the various factors that might (1) discourage or disincentivize reporting such grievances and (2) encourage or incentivize reporting. Lastly, we asked about the redress process as it is currently understood—specifically, what recourse do service members have when trying to resolve racial grievances?

Figure A.1. Four Focus Areas of SME Panel Discussions

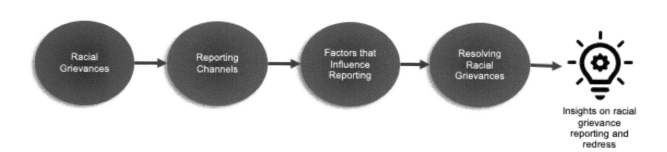

Our SME panelists offered diverse examples of racial grievances that can occur in the military. Even though DoD and DAF policy states that racial grievances be addressed at the lowest level (AFI 26-2710, 2020), we found a variety of examples of racial grievances at every level of the military (i.e., institutional, unit, and individual). Figure A.2 outlines examples of racial grievances that could occur at each level.

Figure A.2. Examples of Possible Racial Grievances in the Military

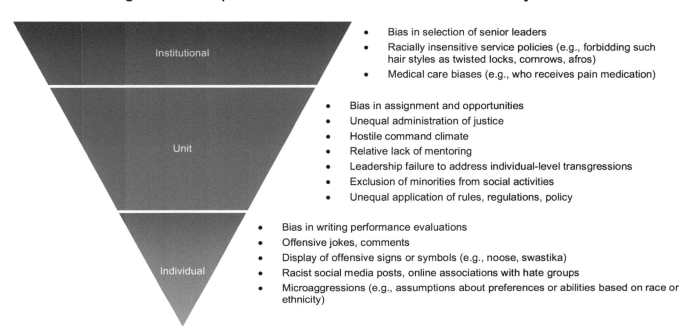

The highest levels of the organization should be actively detecting and addressing institutional issues that have a broad impact across the military system. Senior leaders also need to monitor and potentially address issues within specific units, particularly when the unit leaders are a source of the problem. At the lowest organizational level are problems with the behavior of individuals who do not hold leadership positions. For effective resolution of racial grievances, the military's system needs to be able to identify when it is appropriate to respond to complaints at the individual level and when individual behaviors are merely a symptom or manifestation of wider problems that must be resolved at the unit or institutional level. Preventive efforts should also be located at each organizational level.

Panelists discussed the intersectionality of race and gender and explained the need to properly interpret discrimination experiences that might relate to different aspects of one's identity. Participants described difficult situations in which the aggrieved was targeted along both axes but felt pressure to choose between the two, either labeling the incident as race related or gender related.

We asked panel participants to consider military reporting channels as well as channels external to the military that might play a role in how service members express concern. We observed a high degree of out-the-box suggestions about how aggrieved individuals might choose to report—both internally and externally. The following is a broad snapshot of these conventional and unconventional reporting channels:

- Internal reporting channels:
    - The transgressor
    - Chain of command
    - IG
    - Board for Correction of Military Records
    - Legal counsel and/or judge advocates
    - Unit and organizational climate surveys
    - Unit leadership comment boxes
    - Union: American Federation of Government Employees (civilian wage[21])
    - Peers or mentors
    - Counselors or chaplains
- External reporting channels:
    - Social media
    - Journalists
    - Opinion editorials, blogs
    - Professional association or advocacy group
    - Congressional complaint
    - Special task force or commission
    - Peers
    - Family members.

According to the panelists, the following factors might prevent or discourage reporting:

- "A presumption that 'that's just how the military is'"
- "Unaware of/no access to reporting channels"
- "Command not taking racial issues seriously, seeing it as not their priority/job"
- "Difficult to use (cumbersome reporting process, requires a long written statement)"
- "Lack of trust that the complaint will be handled appropriately or at all"
- "Fear identity would be career ender: kicked off deployment, can't move, could be trapped with offender. Because complaint must be investigated, interviews could reveal ID."
- "Family doesn't want you to risk livelihood/benefits by raising issues/being disruptor"
- "Belief in teamwork, unit cohesion can conflict with notion of reporting"
- "As member of the team, want to brush it off because want the team to look good, team to win, even if shouldn't brush it off"

---

[21] The GS classification and pay system covers most civilian white-collar federal employees (about 1.5 million worldwide) in professional, technical, administrative, and clerical positions.

- "Fear of consequences for reporting: Not promoted, not choice assignment, not choice job; Performance report: can be marked down on teamwork, 'ratting out' your team; Exclusion of a common award: not considered for team award"
- "Lack of support from peers, peer alienation following a report (people afraid to talk, be friends, walking on [thin] ice)"
- "Culture: 'old guard,' old White men, may assume not open to hearing you"
- "Process not quick, easy, time/effort it takes"
- "Outcome not what people were looking for (e.g., too mild, looking for restoration but there was only retribution)"
- "Reputation: perceived as whiny, squeaky wheel, sensitive, 'stink' lingers just for reporting"
- "Active/guard/reserve: can influence connections you have, bias against reserve/guard so might be more dismissive against them"
- "Fear group will ostracize you for being too sensitive"
- "People in multiple categories (intersectional): unsure if racial or choose what path to report (e.g., women of color). Behaviors may span a variety of offenses."
- "Prevalence of a go along to get along mentality"
- "If another rule was violated: being accused of making a report to get out of getting in trouble for your own offense/misconduct."

Participants discussed the potential for those who report to be branded in unfavorable ways that, in addition to affecting their career, might also create an *artificial identity* that they are unable to shed or escape. Specifically, once one reports a racial grievance, others might brand the reporter as a disruptor, a complainer, weak, a "reverse-racist," soft, a poor performer, disgruntled, and more. According to panel participants, these descriptions carry negative connotations that can erode bonds, working relationships, and personal connections forged with fellow service members. For example, one SME panelist noted that individuals who speak out against conscious and unconscious bias could be labeled as "agitators." Also relevant is the primacy of the "mission-first" dictum that suggests any alteration to the status quo is a disruptive, negative one. Along these lines, those who report are often seen as antagonists who erode mission effectiveness rather than protagonists who enhance unit cohesion. Worse yet, such labeling could go beyond mere terminology or descriptions to rebrand the service member's identity, affecting their professional trajectory, peer relationships, and quality of life. For some, this irreversible branding might be worse than the racial offense itself.

Conversely, the following factors might encourage reporting:

- "Inclusive climate and culture, then problem is so far out of the norm it is more likely to be reported"
- "Ombudsmen available to explain process, forms, where to go, who to talk to"
- "Organizational culture seen as receptive: leaders open, shows that reporting is a good things and we need to fix it, vs. don't believe it, says complaint is invalid"
- "A group of people having a desire and mechanism for change"
- "Having seen complaint process work for someone else"

- "After putting up with it for a long time, now getting out of the service willing to 'throw the grenade' [meaning there is less risk of retaliation if no longer under the chain of command]"
- "Leadership makes strong open statement about tolerance, incidents, and follow up with actions (e.g., rename confederate officer named buildings)"
- "Tipping points: really aggrieved, and maybe also those with less to lose"
- "Anonymity, as in the sexual assault process"
- "Mentor or friend to tell it to, can give boost to report"
- "Awareness of options: posted on website, unit bulletin board"
- "Racial representation in leadership"
- "Removal of commanders who do not execute on DEOCS [Defense Organizational Climate Surveys], etc."
- "Peer, supervisor, first sergeant encouragement to report"
- "Desire for retribution"
- "Sense of moral duty to report"
- "Mandate to report"
- "Expectation of improved quality of life"
- "Others have also complained about that person/pattern previously."

The variety of factors influencing reporting spanned the gamut. While culture and command climate can discourage reporting, climate can also work in constructive and influential ways that encourage complainants to report. Whereas pressure might exist to be tough, resilient, and therefore tolerant of aberrant behavior, commanders and leaders who set the opposite tone can facilitate reporting. This goes beyond publicly embracing a zero tolerance policy. Commanders can groom empathetic and culturally compassionate airmen, syncing their DEI objectives with those external to the military. One panelist, a researcher, reflected on the zeitgeist, suggesting that a "culture of sensitivity" might encourage reporting in the military and counter the "culture of toughness" that bisects the services.

The composition of the command climate has a huge bearing on whether or not a complainant decides to report, with more-toxic commands eroding faith in the reporting process. Panelists described key considerations that were likely weighed by the aggrieved: Does the command have a zero tolerance policy? If not, exactly what level of support can I expect from my superiors and peers? Are leaders here reflective and aware or prone to denial and avoidance? Are climate survey results truly representative or somehow skewed? If the command climate is healthy, will my reporting brand it as carcinogenic?

A participant from a military SME panel explained that deciding to report necessitates close scrutiny of the command climate, which involves both leaders and peers. "Preventing racial grievance has a lot to do with zero tolerance policy . . . people come into the military from everywhere and they have to grow in this new environment. Some bring in biases. . . . Positive leadership sets the tone and example, which sets the policy of zero tolerance [and] could help eliminate over time those tendencies when people come in with pre-[conceived] biases."

Panelists frequently cited the impactful nature of military and service-specific cultural norms as affecting the decision to report, either tacitly or explicitly. A question posed by panelists regarding racism, sexism, and other types of discrimination was: Does military culture place an emphasis on asking the appropriate questions and place mission above all else? When thinking about the nexus of norms and reporting, a SME with military experience described military culture as creating a "perpetual problem" and suggested that we consider group differences within the military and subcultures within those groups. From this, we gleaned that military culture might have a vertical or layered construct, with norms at each level having the potential to—either subconsciously or consciously—discourage or encourage reporting (Figure A.3). Although cultural norms may differ at each level, panelists described some norms—such as toughness and resilience—as pervasive, historical in nature, and intertwined with U.S. military identity.

Figure A.3. Top-Down Pervasiveness of Military Cultural Norms

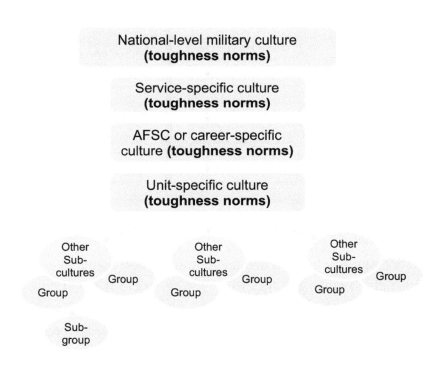

NOTE: AFSC = Air Force Specialty Code (an alphanumeric job identification system).

Lastly, we asked panel participants to comment on the redress process. Our interests were broad and centered on the ways in which discriminatory complaints were received, documented, investigated, and adjudicated. We were also interested in issues related to accountability and how the aggrieved defined and understood *resolution*. Below we include a subset of the many responses we received.

Whittled down to its base components, redress in any environment centers on resolution and the fixing of a problem. However, similar to our first topic, where we asked participants to describe what constitutes a racial grievance, we learned that context figures heavily when discussing resolution. Participants deliberated on what, exactly, resolving a racial grievance might entail and helped to reorient our thinking on the matter. Below are four conceptional reorientations—contextually different ways of thinking about resolving racial grievances that add nuance and value. Figure A.4 depicts the four reorientations and shows the dichotomy between perceived and actual approaches to redress.

Figure A.4. Real Versus Perceived Methods for Considering Redress

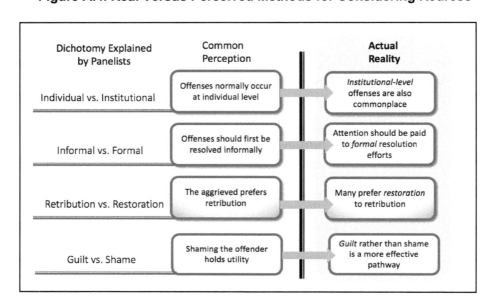

Individual Versus Institutional

Participants across panels suggested that the redress process take into account the tripart dimensionality depicted in Figure A.4. Specifically, if racism or racial offenses can be systemic in nature, condoned or overlooked by senior leaders, or perpetrated by airmen at junior ranks, the resolution process should be equipped to address problems effectively and efficiently at each level. Solutions should be tailored, or mapped, to the type of complaint and the associated actor (i.e., individual, leadership, or institution [which relates to systemic or cultural issues]). Examples of resolution actions tailored to the individual include public or private apologies, relocating the complainant or offender, and dismissal or termination of the offender. Conversely, examples of institutional-level resolution actions are setting standards for fair, equitable, and nondiscriminatory behavior and consequences for failing to meet those standards; pervasive retooling of unconscious bias training; revision of policy, rules, regulations, and guidance; and a public commitment to changing institutional culture. As frequently mentioned by panelists, institutional-level actions are also instrumental in ameliorating service-specific cultural issues.

Informal Versus Formal

Pressure is often applied to resolve racial complaints informally and at the lowest levels rather than elevate issues higher up the chain of command. Panelists described this problem in a myriad of ways. Some suggested that informal resolutions amount to "papering it over" or prescribing an easy, "low-visibility" fix that obviates senior leadership from aggressively addressing the issue and considering whether there are institutional- or systemic-level origins. Under such circumstances, the complainant might be pressured to accept a mild apology, or the offender might be encouraged to explain it away. A participant from a panel of experienced researchers remarked, "I think the fact that incidents are not tracked, no paper trail, encouraged to resolve at lowest level, many things prevent them from getting to the point of formal report . . . . [there is no] way of circling back with the victim to see if they were truly satisfied with the way things were resolved." Another person described uniform personnel who are motivated to not file racial grievance reports so then can "protect their own careers" and receive positive performance evaluations. The ramifications were described to us as serious. In an age in which proper messaging and optics matter, informal resolutions send a signal that the military sees race-related issues as less serious and more trivial. Informal resolutions can also hide harmful offender behavior.

Retribution Versus Restoration

Although it is a common assumption that the aggrieved seeks retribution for offenses occurred, the panelists brought to our attention that there might be a desire for restorative justice. The key is to pay close attention to the preferences of the aggrieved. For example, if they wanted what was denied to them restored (e.g., higher rank or career advancement), they will not be satisfied with command solely punishing the offender. If they wanted the offender held accountable or to prevent similar race-related incidents from happening to others, they will not be satisfied with a higher rank or better position. Law professor Jac Armstrong, when describing what he calls the *restorative-retributive dichotomy*, admits to the absence of a definition of *restorative justice* but claims that it "should seek to maximize stakeholder participation (with stakeholders being defined as the victim, the offender, and the community, but excluding the state) and that offenses create obligations between the offender and the victim, with outcomes measured through victim satisfaction" (Armstrong, 2014, p. 363; Marshall, 1999). R. A. Duff, professor of law, suggests that offenders should suffer punishment for their offenses but that the objective of the punishment should be to achieve restoration, meaning that "the offender's normative relationship towards both the community and the victims must be restored, returning to its previous equilibrium" (Duff, 2002, p. 101).

Guilt Versus Shame

The guilt-shame dichotomy works in similar fashion to the retribution-restoration dichotomy in that exercising one set of behaviors over another can yield drastically different and, possibly, unintended consequences. A nonmilitary panelist commenting on the redress process explained that if someone feels guilty, they are interested in doing better next time. However, if they feel ashamed, they usually regress or stay away from restorative action. Therefore, shaming perpetrators might yield counterproductive results for the victims in the long term. Brown et al. (2008, p. 76) states,

> There is some evidence that when people perceive themselves to have behaved illegitimately, they can experience guilt or shame (or both) and that, depending on which predominates, rather different outcomes can occur. Although the evidence is not unequivocal, guilt seems to be more closely connected to prosocial orientations and shame to stronger negative self-evaluations, reputational concerns, and various kinds of avoidance behavior.

Participants suggested that we consider the importance of "deciders" or "action takers" being methodical and careful in their assessment and adjudication process and obtaining a nuanced appreciation or the context and specific circumstances before determining a course of action. Hastily rendered decisions run the risk of missing compounding actions by the offender. Key questions to ask include the following: Is this person disrespectful in other ways, too? Have they been repeatedly warned? Were there previous complaints (even if unproven)? Did the investigator potentially have a conflict of interest or bias?

Other actions that warrant concern are the hurried dismissal of complaints, attempts to explain away offensive behavior, and attempts to undermine the complainant. Panel participants flagged for us common refrains that crop up during the redress process, such as "They didn't mean it," "Are you instead disgruntled for a different reason?" "You're being too sensitive," "That didn't really happen," "You can't prove that," and "You're trying to game the system." In such circumstances, the undergirding sentiment tends to be disbelief and mistrust of the complainant, especially if evidence is incomplete, indecipherable, or hard to produce. Even if there is no direct evidence of the wrongdoing to substantiate a complaint, when someone comes forward, leaders should take actions to fulsomely assess the offender and, when applicable, address the unit culture, attitudes, and behavior.

Lastly, we asked panel participants to comment on the redress process. Our interests were broad and centered on the ways in which discriminatory complaints were received, documented, investigated, and adjudicated. We were also interested in issues related to accountability and how the aggrieved defined and understood *resolution*. Below we share a subset of our SME panel's responses on ideal actions to resolve racial grievances:

- Individual-level resolution actions:
  - "Public or private apology"
  - "Move complainant or offender"
  - "Restoration of rank, rights, privileges, anything that was unjustly taken away"
  - "Dismiss offender, they lose retirement benefits"
  - "Financial compensation for complainant"

- "Restorative justice/arbitration/mediation: bring the parties together to work out a resolution"
- "Feeling heard—grievance acknowledged by leadership, even if not by offender themselves"

- Institutional-level resolution actions:
  - "Information campaign (not just the provision of data with context)"
  - "Unconscious bias addressed in training"
  - "Not just band-aids: culture change writ large, top down full recognition of reality of racism and impacts, [and improvement of the cultural diversity] of recruiting pools and locations of military bases"
  - "Revision of policy, rules, regulations, guidance"
  - "Set standards and consequences for failure to meet standards"
  - "Transparency about the outcome, which can give comfort to complainant, encourage reporting too, and be deterrent for future offenders"
  - "Better monitoring of social media, where the comments are taking place"
  - "Civilian and military top down zero tolerance for racism [which might prevent feelings that the chain of command condones or will not reveal racism is condoned] can cut down sense that it's condoned [or that people will not be held accountable for racist acts]"
  - "Referral of problem to third party independent external regulatory committee (outside the military) to decide and implement redress."

The following are high-level takeaways consolidated into a list for easy reference:

- The reporting and redress system is designed to be reactive—to address individual and interpersonal racial grievances and to address wrongs.
- The system has to be designed to address a wide variety of issues that could trigger a racial grievance.
- Although reporting through the chain of command is framed as the de facto method for military service members, it might be off-limits to the aggrieved and mandate a different approach.
- Indirect reporting channels, or working through intermediaries (e.g., chaplains or mentors), can offer comfort, ease, and confidentiality.
- Social media is a game changer, as people can and do share information that can go viral and/or directly reach higher levels of authority.
- Choosing to report might saddle complainants with an unfavorable, contrived identity.
- A constructive command climate with consistent, proactive messaging is a game changer for the aggrieved.
- Military cultural norms that emphasize toughness and resilience might inadvertently discourage reporting.
- The influence of the societal zeitgeist (external to the military) can encourage reporting.
- Actions need to map to the type of complaint.
- The pressure to resolve racial complaints informally and at the lowest level contributes to a lack of visibility on race-related problems and an inability to adequately address them.

- Be mindful of the difference between restorative and retribution actions. Find out what the complainant was seeking by reporting.
- Deciders need to have a nuanced appreciation of the context and circumstances before determining a course of action.
- Record, track, and use the data on complaints, decisions, actions taken, and short- and long-term outcomes.
- The system should be proactive. Do not just wait for someone to file a report but ensure there are top-down, multilevel efforts seeking to understand, find patterns, solicit input, and monitor how grievances are handled.

Appendix B. History of Race in the U.S. Military

This appendix provides a historical timeline of racial and ethnic disparities in opportunity and discipline that complements Chapter 2 of this report.[22]

American Revolution (1775–1783)

Policies and practices that sought to exclude or limit Black people from military service began with the fight for U.S. independence and the creation of a standing army by the Second Continental Congress on June 14, 1775.[23] Historians estimate that approximately 5,000 Black soldiers picked up arms to fight for independence. Yet, despite the Black soldiers' service, Congress enacted the Militia Act of 1792, which barred racial and ethnic minorities from fighting for the U.S. Army (DEOMI, 2002). In 1798, the U.S. Marine Corps adopted a policy forbidding the enlistment of "Negroes, Mulattos, and Indians"; this policy and practice remained in place until 1942 (DEOMI, 2002, p. 2).

War of 1812 (1812–1815)

In policy and practice, racial and ethnic minorities continued to serve during personnel and manpower shortages in the War of 1812. Free and enslaved Black people were allowed to serve as soldiers and sailors, but the U.S. Marine Corps policy remained intact.

Both the U.S. and British militaries offered paths to freedom for the enslaved, but when the Treaty of Ghent ended hostilities between the two nations at the end of 1814, they also agreed that "all possessions" taken during the war "shall be restored without delay," including "any slaves or other private property" (Lousin, 2014, p. 1). Ultimately, the British compensated the enslavers $1,204,960 in lieu of returning any enslaved people who enlisted in the British Armed Forces. In the United States, despite fighting bravely, enslaved Black people were denied their promises of freedom and returned to the enslavers.

At the end of the hostilities, the U.S. Army issued an order on February 18, 1820, that stated, "No Negro or Mulatto will be received as a recruit of the Army" (Foner, 1974, p. 27). One year later, the Army instituted regulations restricting enlistment to "all free White male persons," which excluded Black people from service but permitted them to work as servants and laborers.

---

[22] This appendix and Esposito and Gregory (2021) share some common descriptions of this historical context, as they were written concurrently by the same authors for a similar audience.

[23] The U.S. Army reversed the policy set by Adjutant General Horatio Lloyd Gates only when he directed recruiters that "You are not to Enlist any Deserter from the Ministerial Army, nor any Stroller, Negro or Vagabond." See National Archives, undated-b.

In 1839, the Secretary of the Navy limited Black people to 5 percent of the overall service—a policy that remained in effect until the Civil War (Foner, 1974).

The Civil War (1861–1865)

Facing mass casualties and a declining number of White volunteers, Congress passed the Militia Act of 1862, making it legal for "Negro men to enlist in the United States Army for the purpose of constructing intrenchments, or performing camp service or any other labor, or any military or naval service for which they may be found competent" (U.S. Congress, 1862, Chapters 200–201). The Union Army organized Black troops into segregated units commanded by White officers, but these units were used limitedly in combat. Black troops were subjected to discriminatory practices, including lower pay ($7 versus $13 per month for Black and White members, respectively), poor equipment, and conditions and medical care resulting in a nearly 40 percent higher mortality rate than their White counterparts. Initially, the Navy lifted its 5 percent quota on Black sailors but restricted them to positions of servants, cooks, assistant gunners, and powder boys and excluded enlisted Black sailors from petty officer and commissioned officer ranks (DEOMI, 2002).

By the end of the Civil War, approximately 179,000 Black men served in the Union Army (making up 10 percent of the army), and another 19,000 served in the Navy (Bryan, undated). Approximately 90,000 served with the Confederate Army (DEOMI, 2002). Over the course of the war, nearly 40,000 Black soldiers died; approximately 30,000 perished from infection or disease (National Archives, undated-a).

World War I (1914–1918)

By the time the United States entered the war against Germany in April 1917, Black people were barred from the Marine Corps and the Army Aviation Corps and were limited to such menial positions as messmen, cooks, and coal heavers in the Navy (DEOMI, 2002). U.S. Army policy allowed Black soldiers to serve in any position, but, in practice, the majority were placed in service or supply regiments, serving as stevedores, drivers, engineers, and laborers.

Facing backlash from the Black community, the War Department ultimately created the 92nd and 93rd Divisions, both primarily Black combat units (Bryan, undated). These two combat divisions had vastly different experiences in theater. The 92nd Division served under White U.S. officers in France, was subject to racist treatment (e.g., it was referred to as the "Rapist Division" by the division commander), and suffered low morale (DEOMI, 2002). Furthermore, the unit's combat record was mixed at best, as it failed to meet a crucial mission during the Meuse-Argonne offensive (Bryan, undated). With strong correlation between morale and battlefield performance, the operational risks of discrimination had significant negative outcomes for the war effort (Lyall, 2020). Afterward, regardless of how well the unit actually performed on the battlefield, it was never able to overcome the slander from prejudiced officers (Lyall, 2020).

The 93rd Division fought under the French and had a stellar combat record, with 170 individual soldiers and three infantry regiments earning France's highest military award, the Croix de Guerre (DEOMI, 2002). The French military command fundamentally had no issues with Black service members; it found them courageous and honorable and treated them more humanely than did their U.S. counterparts. Isabel Wilkerson discusses this dynamic in her book *Caste*—French leadership referred to the African American soldiers with respect to their White leadership as follows: "even though the choicest with respect to physique and morals . . . we cannot deal with them on same plane as the white American officers without deeply wounding the latter" (Wilkerson, 2020, p. 225). Despite the 93rd Division's serving courageously and capably in combat and its achievements, military authorities spent the next 30 years citing the failure of the 92nd as evidence of the inadequacy of Black soldiers in combat (Bryan, undated).

Post–World War I

Nearly 400,000 Black soldiers served during the Great War; 200,000 deployed overseas, and 42,000 served in the combat units (Roza, 2020). Black soldiers returned home to widespread Jim Crow laws and increasing racial tensions that boiled over by the summer of 1919.[24] Among the victims of the Red Summer, at least ten were war veterans, some of whom were lynched while in uniform.[25]

In 1925, the U.S. Army War College (USAWC) submitted a report titled *Employment of Negro Man Power in War* to the Chief of Staff of the Army, with the recommendation "that it be accepted as the War Department policy in handling this problem" (Ely, 1925). The study describes the problem as follows:

> The War Department had no pre-determined and sound plan for the use of Negro troops at the beginning of the World War. It had no adequate defense against political and racial pressure and was forced to organize negro combat divisions and commission unqualified negro officers. (Ely, 1925)

The USAWC recommended employment of the following strategy:

> If it can be shown that the Negro is given an equal opportunity with the White man to qualify for commissioned grades, and that only his lack of qualifications prevent his commission in the higher grades or in combat units, then social and political demands of the administration can be resisted. (Kamarck, 2019, p. 4)

---

[24] Jim Crow laws were a collection of state and local statutes that legalized racial segregation. Named after a Black minstrel show character, the laws—which existed for about 100 years, from the post–Civil War era until 1968—were meant to marginalize African Americans by denying them the right to vote, hold jobs, and get an education, as well as other opportunities. Those who attempted to defy Jim Crow laws often faced arrest, fines, jail sentences, violence, and death. See "Jim Crow Laws" (2018).

[25] Known as the *Red Summer*, violent anti-Black riots erupted in 26 cities across the United States, and lynchings increased from 64 in 1918 to 83 in 1919. See National WWI Museum and Memorial (undated) and Bryan (undated).

Regretfully, this illusion of equity and inclusion perhaps has left a legacy that is manifest in the inadequate attempts to make improvements today.

This strategy employs the appearance of EO in policy while continuing to practice denial of opportunity for racial and ethnic minorities to serve in combat units and as officers.[26] The USAWC report emphasized that Black members are a class "from which we *cannot* expect to *draw leadership* material" and "*all* officers, *without exception*, agree that the Negro lacks initiative, displays little to no leadership, and cannot accept responsibility" (Ely, 1925) (emphasis added). Yet, Black service members would have to rely on these very officers to give them a fair and equitable opportunity to succeed.

Although this report was submitted to the Army Chief of Staff on October 30, 1925, its impact on military accession and promotion of Black people and other minority groups continues to this day, as reflected in these groups' notable absence in combat career fields and the senior officer corps.

World War II (1939–1945)

During World War II, military leaders continued to believe that racial minorities were unfit for combat or leadership positions and continued to relegate them to segregated labor and service units. The Army upheld a 10 percent quota for Black recruits, and, by 1941, Black soldiers in the Army accounted for 5 percent of the Infantry and less than 2 percent of the Air Corps (Kamarck, 2019). In contrast, Black soldiers accounted for 15 percent of the Quartermaster Corps. Similarly, 2 percent of the Navy was Black, and nearly all of them, except for six seamen, were relegated to the Steward Mate Corps; none were officers. However, immense political pressure and mobilization requirements soon generated changes to defense policies, despite considerable resistance from military leaders.

Between 1939 and 1940, Congress enacted three laws that significantly affected Black participation in the Army Air Corps (Osur, 2000):

- *The Civilian Pilot Training Act* established the Civilian Pilot Training Program to create a reserve of civilian pilots who could be quickly mobilized for war. Black citizens participated in this program at several sites, including the Tuskegee Institute.
- *Public Law 18* provided for the large-scale expansion of the Army Air Corps and stipulated that one of the civilian contract schools had to be designated for the training of racial minorities.

---

[26] While stating that a Black person "should be given a fair opportunity to perform the tasks in war for which he is qualified or may qualify," the same report weaponized prevailing racial stereotypes to exclude Black people from combat or leadership (Ely, 1925). The racial stereotypes found in this memorandum that were used as justification for the exclusion of Black people from military service include Black people being mentally inferior, inherently weak in character, and *rank cowardice* (meaning that they had extreme fear of the unknown).

- In response to resistance to the two previous laws, Congress passed the Selective Training and Service Act of 1940, which prohibited discrimination on the basis of race and color.[27]

However, military leaders continued to resist, taking the position that although Congress required them to *train* Black service members, they were not required to *employ* them. Furthermore, because no provisions were made to create Black units in the Air Corps, there was no way to use them. General Henry "Hap" Arnold, Chief of the Air Corps, reiterated this position and added, "negro pilots cannot be used in our present Air Corps units since this would result in having negro officers serving over White enlisted men, creating an impossible social problem" (Osur, 2000).

By early 1941, the Roosevelt administration, under increasing public pressure, ordered the War Department to create a Black flying unit. On January 16, 1941, the War Department announced the creation of the 99th Pursuit Squadron, and, in March 1941, the Air Corps established the all-Black 99th Pursuit Squadron, known today as the Tuskegee Airmen. On June 25, 1941, President Franklin D. Roosevelt issued Executive Order 8802, establishing the Fair Employment Practices Commission, and created a policy of nondiscrimination in all branches of the Armed Service (Executive Order 8802, 1941). The following year, under pressure from the Roosevelt administration, the Marine Corps began recruiting Black citizens into segregated units for the first time since 1798; none were officers (Osur, 2000).

Post–World War II Military Integration

On July 26, 1948, President Harry Truman issued Executive Order 9981 to end segregation in the military and ordered the full integration of all the services (Kamarck, 2019). The order declared, "there shall be equality of treatment and opportunity for all persons in the armed forces without regard to race, color, religion, or national origin" (Executive Order 9981, 1948, p. 1) The order also established the President's Committee on Equality of Treatment and Opportunity in the Armed Services, led by Charles Fahy, to examine the potential impact of integration on military efficiencies.

The committee's report, released in 1950, found that "existing segregation policies were contributing to inefficiencies through unfilled billets, training backlogs, and less capable units" (Kamarck, 2019, p. 4). Despite considerable resistance to the executive order from military

---

[27] 50 U.S.C. §3805 states,

> The selection of persons for training and service under section 3803 of this title shall be made in an impartial manner, under such rules and regulations as the President may prescribe, from the persons who are liable for such training and service and who at the time of selection are registered and classified, but not deferred or exempted: Provided, That in the selection of persons for training and service under this chapter, and in the interpretation and execution of the provisions of this chapter, there shall be no discrimination against any person on account of race or color.

See 50 U.S.C. §3805, which was repealed by Pub. L. 91-124, Section 2 (1969).

leaders, the manpower needs of the Korean War (1950–1953) catalyzed racial integration in the services, and, by 1954, then–Secretary of Defense Charles Erwin Wilson announced that the last all-Black active duty unit had been abolished (Gropman, 1985, p. 145).

Appendix C. Comparative Analysis of Definitions of Diversity Across the Services

DoD released a definition of *diversity* on February 5, 2009, which is specified in DoDD 1020.02E (2018). This overarching definition is very broad and includes characteristics and attributes (e.g., gender, religion, race, or ethnicity) that may or may not be legally protected. Although it defines diversity, it lacks a statement on DoD's organizational diversity goals, which would provide guidance on how to operationalize diversity. Each service has also developed its own diversity definition that aligns with the DoD definition and includes further language on organizational diversity goals. Some services, such as the Marine Corps, have implicitly defined diversity. These service-specific definitions are wide ranging and representative. They connect diversity to mission readiness and the service's core values. For example, the Army notes how diversity among its members can promote cultural competence internationally. The Army, Navy, and Marine Corps definitions also indicate how the inclusion of a combination of diverse attributes make the service stronger as a whole.

The following list shows the varying definitions of diversity in DoD and service policies:

- **DoD:** Diversity is "[a]ll the different characteristics and attributes of DoD's total force, which are consistent with DoD's core values, integral to overall readiness and mission accomplishment, and reflective of the Nation we serve" (DoDD 1020.02E, 2018, p. 13).
- **DAF:** The DAF broadly defines diversity as "a composite of individual characteristics, experiences, and abilities consistent with the Air Force Core Values and the Air Force Mission. Air Force diversity includes but is not limited to: personal life experiences, geographic and socioeconomic backgrounds, cultural knowledge, educational background, work experience, language abilities, physical abilities, philosophical and spiritual perspectives, age, race, ethnicity, and gender" (AFI 36-7001, 2019, p. 3).
- **Army:** The Army defines diversity as "the different attributes, experiences, and backgrounds of our Soldiers, Civilians and Family Members that further enhance our global capabilities and contribute to an adaptive, culturally astute Army" (U.S. Army Regulation 600-100, 2017, p. 32).
- **Navy:** Per the U.S. Navy's *Inclusion and Diversity: Goals and Objectives* report: "Diversity means all the different characteristics and attributes of our Navy Team, which are consistent with Navy core values, integral to overall readiness and mission accomplishment and reflective of the Nation we serve" (U.S. Navy, 2020, p. 6). However, the Military Leadership Diversity Commission presented a more robust definition of diversity: "the term diversity encompasses not only the traditional categories of race, religion, age, gender, national origin, but also all the different characteristics and attributes of individuals that enhance the mission readiness of the Department of the Navy and strengthen the capabilities of our Total Force: Sailors, Marines, Government Civilians, and Contractors" (Military Leadership Diversity Commission, 2011, p. 13).

- **Marine Corps:** The Marine Corps Prohibited Activities and Conduct Prevention and Response Policy defines cultural diversity as "a condition in a group of people or organization brought about by the gender, religion, racial, cultural, and social differences that the individuals naturally bring to the group or organization" (Marine Corps Order 5354.1F, 2021, p. A-5).

Abbreviations

AFI	Air Force Instruction
AFOSI	Air Force Office of Special Investigations
ADC	Area Defense Counsel
CDI	Commander Directed Investigation
DAF	U.S. Department of the Air Force
DAFGM	Department of the Air Force Guidance Memorandum
DEI	diversity, equity, and inclusion
DEOMI	Defense Equal Opportunity Management Institute
DoD	U.S. Department of Defense
DoDD	Department of Defense Directive
DoDI	Department of Defense Instruction
DOTMLPF	doctrine, organization, training, materiel, leadership and education, personnel, and facilities
EO	equal opportunity
EEO	equal employment opportunity
GCMCA	General Court-Martial Convening Authority
GS	General Schedule
HHQ	higher headquarters
IG	Inspector General
IT	information technology
MAJCOM	major command
MEO	military equal opportunity
NJP	nonjudicial punishment
PAF	Project AIR FORCE
RDR	Racial Disparity Review
SAF	Secretary of the Air Force
SARC	sexual assault response coordinator
SJA	Office of the Staff Judge Advocate
SME	subject-matter expert
UCMJ	Uniform Code of Military Justice
UIF	unfavorable information file
USAWC	U.S. Army War College

Bibliography

AFI—*See* Air Force Instruction.

Air Command and Staff College, *AU-2 Guidelines for Command: A Handbook on the Leadership of Airmen for Air Force Squadron Commanders*, Air University Press, March 2015.

Air Force Doctrine Publication 1, *The Air Force*, Department of the Air Force, March 10, 2021.

Air Force Doctrine Volume 2, *Leadership*, Department of the Air Force, August 8, 2015.

Air Force Instruction 1-2, *Commander's Responsibilities*, Secretary of the Air Force, May 8, 2014.

Air Force Instruction 10-601, *Operational Capability Requirements Development*, Secretary of the Air Force, November 6, 2013.

Air Force Instruction 36-7001, *Diversity and Inclusion*, Secretary of the Air Force, February 19, 2019.

Air Force Instruction 90-301, *Inspector General Complaints Resolution*, Secretary of the Air Force, December 28, 2018, change 1, September 30, 2020.

Air Force Policy Directive 36-27, *Equal Opportunity (EO)*, Secretary of the Air Force, March 18, 2019.

Armstrong, Jac, "Rethinking the Restorative-Retribution Dichotomy: Is Reconciliation Enough?" *Contemporary Justice Review*, Vol. 17, No. 3, 2014.

Baldus, David C., Catherine M. Grosso, George Woodworth, and Richard Newell, "Racial Discrimination in the Administration of the Death Penalty: The Experience of the United States Armed Forces (1984–2005)," *Journal of Criminal Law & Criminology*, Vol. 101, No. 4, 2012.

"Black Soldiers Timeline," *Encyclopedia of Slavery and Abolition*, American Abolitionists and Antislavery Activists: Conscience of the Nation, webpage, updated April 4, 2021. As of September 1, 2021:
http://www.americanabolitionists.com/black-soldiers---timeline.html#Black%20Soldiers%20Timeline%201638-1862

Brown, Rupert, Roberto González, Hanna Zagefka, Jorge Manzi, and Sabina Cehajic-Clancy, "Nuestra Culpa: Collective Guilt and Shame as Predictors of Reparation for Historical Wrongdoing," *Journal of Personality and Social Psychology*, Vol. 94, No. 1, 2008.

Bryan, Jami, L., "Fighting for Respect: African-American Soldiers in WWI," National Museum of the United States Army, webpage, undated. As of September 1, 2021: https://armyhistory.org/fighting-for-respect-african-american-soldiers-in-wwi/

Chairman of the Joint Chiefs of Staff Instruction 5123.01H, *Charter of the Joint Requirements Oversight Council and Implementation of the Joint Capabilities Integration and Development System*, Joint Staff, October 30, 2021.

Christensen, Don, and Yelena Tsilker, *Racial Disparities in Military Justice: Findings of Substantial and Persistent Racial Disparities Within the United States Military Justice System*, Protect Our Defenders, May 5, 2017.

Cohen, Rachel S., "Brown: 'Shame on Us' if Military Diversity Efforts Falter," *Air & Space Forces*, December 22, 2020.

DAF IG—see U.S. Department of the Air Force Inspector General.

Death Penalty Information Center, "Racial Disparity in the Military Death Penalty," webpage, undated. As of August 14, 2023: https://deathpenaltyinfo.org/state-and-federal-info/ military/racial-disparity-in-the-military-death-penalty

Defense Equal Opportunity Management Institute, *Historical Overview of Racism in the Military*, Special Series Pamphlet 02-1, February 2002.

DEOMI—*See* Defense Equal Opportunity Management Institute.

Department of Defense Directive 1020.02E, *Diversity Management and Equal Opportunity in the DoD*, incorporating change 2, June 1, 2018.

Department of Defense Directive 1440.1, *The DoD Civilian Equal Employment Opportunity (EEO) Program*, change 3, April 17, 1992.

Department of Defense Instruction 1020.03, *Harassment Prevention and Response in the Armed Forces*, change 2, December 29, 2020.

Department of Defense Instruction 1020.05, *DoD Diversity and Inclusion Management Program*, September 9, 2020.

Department of Defense Instruction 1350.02, *DoD Military Equal Opportunity Program*, change 1, September 4, 2020.

Department of Defense Instruction 1350.03, *Affirmative Action Planning and Assessment Process*, February 29, 1988.

Department of the Air Force Guidance Memorandum (DAFGM) to the Department of the Air Force Instruction (DAFI) 36-2710, *Equal Opportunity Program*, September 2, 2021.

DoD—*See* U.S. Department of Defense.

DoDI—*See* Department of Defense Instruction.

Duff, R. A., "Restorative Punishment and Punitive Restoration," in Lode Walgrave, ed., *Restorative Justice and the Law*, Willan Publishing, 2002.

Ely, H. F., "Employment of Negro Man Power in War," memorandum for the Chief of Staff A.C. 27-25, November 19, 1925.

Esposito, Albert M., and Chong H. Gregory, "Great Power Conflict: Competitive Advantage Through Diverse Leadership," Air Force Fellows Program, Air University, April 25, 2021.

Executive Order 8802, "Prohibition of Discrimination in the Defense Industry," Executive Office of the President, June 25, 1941.

Executive Order 9981, "Establishing the President's Committee on Equality of Treatment and Opportunity in the Armed Services," Executive Office of the President, July 26, 1948.

Executive Order 13950, "Combatting Race and Sex Stereotyping," Executive Office of the President, September 22, 2020.

Executive Order 13985, "Advancing Racial Equity and Support for Underserved Communities Through the Federal Government," Executive Office of the President, January 20, 2021.

Foner, Jack D., *Blacks and the Military in American History*, Praeger Publishers, 1974.

Fedrigo, John A., "Policies to Improve Accountability Within a Diverse Department of the Air Force," memorandum, Office of the Assistant Secretary, Manpower and Reserve Affairs, December 21, 2020.

Goldfein, Dave, *CSAF Focus Area: The Beating Heart of the Air Force . . . Squadrons!* U.S. Air Force, August 2016.

Gropman, Alan L., *The Air Force Integrates 1945–1964*, Office of Air Force History, 1985.

Halvorsen, Howard E., "Air Force History: CMSAF #4 Thomas N. Barnes," Tinker Air Force Base, webpage, February 13, 2017. As of August 31, 2021:
https://www.tinker.af.mil/News/Article-Display/Article/1082072/
air-force-history-cmsaf-4-thomas-n-barnes/

Jeffrey, James, "Remembering the Black Soldiers Executed After Houston's 1917 Race Riot," *The World*, February 1, 2018.

"Jim Crow Laws," History.com, February 28, 2018. As of September 1, 2021:
https://www.history.com/topics/early-20th-century-us/
jim-crow-laws?form=MY01SV&OCID=MY01SV

Judge Advocate General's School, *The Military Commander and the Law*, 15th edition, Air University Press, 2019.

Kamarck, Kristy N., *Diversity, Inclusion, and Equal Opportunity in the Armed Services: Background and Issues for Congress*, Congressional Research Service, R44321, June 5, 2019.

Lousin, Ann M., "On the Bicentennial of the Treaty of Ghent, Reflecting on Its Slavery Clauses" *Chicago Daily Law Bulletin*, Vol. 160, No. 252, 2014.

Lyall, Jason, *Divided Armies: Inequality and Battlefield Performance in Modern War*, Princeton University Press, 2020.

MacGregor, Morris J., Jr., *Integration of the Armed Forces 1940–1965*, Center of Military History, United States Army, 1981.

Marine Corps Order 5354.1F, "Marine Corps Prohibited Activities and Conduct (PAC) Prevention and Response Policy," Department of the Navy, April 21, 2021.

Marshall, Tony F., *Restorative Justice: An Overview*, Home Office Research Development and Statistics Directorate, 1999.

MacLaury, Judson, "The Federal Government and Negro Workers Under President Woodrow Wilson," U.S. Department of Labor, Annual Meeting of the Society for History in the Federal Government, March 16, 2000.

Military Leadership Diversity Commission, "Defining Diversity," decision paper #5, February 2011.

National Archives, "Black Soldiers in the U.S. Military During the Civil War," Educator's Resources, webpage, undated-a. As of September 1, 2021: https://www.archives.gov/education/lessons/blacks-civil-war

National Archives, "Horatio Gates Papers," National Historical Publications and Records Commission, webpage, undated-b. As of September 1, 2021: https://www.archives.gov/nhprc/projects/catalog/horatio-gates

National WWI Museum and Memorial, "Red Summer: The Race Riots of 1919," webpage, undated. As of September 1, 2021: https://www.theworldwar.org/learn/wwi/red-summer

Office of People Analytics, *2015 Workplace and Equal Opportunity Survey of Reserve Component Members*, Defense Technical Information Center, July 2018.

Office of People Analytics, *2017 Workplace and Equal Opportunity Survey of Active Duty Members: Executive Report*, Defense Technical Information Center, August 1, 2019.

Office of Special Investigations, *OSI 2020 Fact Book: Expanding Our Mission Frontier*, Vol. 46, No. 2, 2020.

Office of the Deputy Assistant Secretary of Defense for Military Community and Family Policy, *2019 Demographics: Profile of the Military Community*, U.S. Department of Defense, 2019.

Office of the Secretary of Defense, *Report of the Task Force on the Administration of Military Justice in the Armed Forces*, November 30, 1972.

Osur, Alan M., *Separate and Unequal: Race Relations in the AAF During World War II*, Air Force History and Museums Program, May 2000.

Protect Our Defenders, "Military Justice Overview," webpage, undated. As of August 31, 2021: https://www.protectourdefenders.com/downloads/Military_Justice_Overiew.pdf

Protect Our Defenders, *Federal Lawsuit Reveals Air Force Cover Up: Racial Disparities in Military Justice Part II*, May 2020.

Public Law 91-124, To Amend the Military Selective Service Act of 1967 to Authorize Modifications of the System of Selecting Persons for Induction into the Armed Forces Under This Act, November 26, 1969.

Robinson, Barry K., and Edgar Chen, "Déjà Vu All Over Again: Racial Disparity in the Military Justice System," *Just Security*, September 14, 2020.

Roza, David, "They Were Among the Fiercest American Soldiers in WWI. Here's Why They Were Horribly Mistreated When They Returned Home," *Task and Purpose*, March 11, 2020.

Secretary of the Air Force, Office of the Inspector General Complaints Resolution Directorate, *Commander Directed Investigation (CDI) Guide*, Joint Base Anacostia-Bolling, June 1, 2018.

Stewart, Phil, M. B. Pell, and Joshua Schneyer, "U.S. Troops Battling Racism Report High Barrier to Justice," Reuters, September 15, 2020.

Theus, Lucius, "Air Force Regulation 30-1: Race Relations in the U.S. Air Force." *Commander's Digest*, Vol. 13, No. 4. 1972.

UCMJ—*See* Uniform Code of Military Justice.

Uniform Code of Military Justice, Appendix 2, Section 815, Article 15, Commanding Officer's Non-Judicial Punishment, December 20, 2019.

U.S. 37th Congress Session II, *Militia Act 1862*, July 17, 1862.

U.S. Air Force Guidance Memorandum to Air Force Instruction 36-2710, *Equal Opportunity Program*, Department of the Air Force, September 9, 2020.

U.S. Army Regulation 600-100, *Army Profession and Leadership Policy*, Department of the Army, April 5, 2017.

U.S. Code, Title 10, Section 481, Racial and Ethnic Issues; Gender Issues: Surveys.

U.S. Code, Title 10, Section 9233, Requirement of Exemplary Conduct.

U.S. Code, Title 50, Section 3805, Manner of Selection of Men for Training and Service; Quotas.

U.S. Department of Defense, *Writing Style Guide and Preferred Usage for DoD Issuances*, February 10, 2020.

U.S. Department of Defense, "Department of the Air Force Guidance Memorandum to Department of the Air Force Instruction 36-2710, *Equal Opportunity Program*," July 20, 2023.

U.S. Department of the Air Force Inspector General, "Who We Are," webpage, undated. As of September 1, 2021:
https://www.afinspectorgeneral.af.mil/Who-We-Are/

U.S. Department of the Air Force Inspector General, *Report of Inquiry (S8918P): Independent Racial Disparity Review*, December 2020.

U.S. Department of the Air Force Inspector General, *Assessment Report (S8918P): Independent Racial Disparity Review Six-Month Assessment*, September 2021.

U.S. Department of the Army, *Commander's Equal Opportunity Handbook*, Training Circular No. 26-6, June 23, 2008.

U.S. Government Accountability Office, *Military Justice: DOD and the Coast Guard Need to Improve Their Capabilities to Assess Racial and Gender Disparities*, GAO-19-344, May 2019.

U.S. Equal Employment Opportunity Commission, *Title VII of the Civil Rights Act of 1964*, Public Law 88-352, as amended as it appears in Volume 42 of the United States Code beginning at section 2000e, 1964.

U.S. Equal Employment Opportunity Commission, "Race/Color Discrimination," webpage, undated. As of October 19, 2023:
https://www.eeoc.gov/racecolor-discrimination

U.S. Federal Aviation Authority, JO 7110.65Z, 2023. As of August 31, 2021:
https://www.faa.gov/air_traffic/publications/atpubs/atc_html/chap1_section_2.html

U.S. House Armed Services Committee, "Racial Disparity in the Military Justice System—How to Fix the Culture," hearing before the Subcommittee on Military Personnel, June 16, 2020.

U.S. Navy, *Inclusion and Diversity: Goals and Objectives*, 2020.

Wilkerson, Isabel, *Caste: The Origins of Our Discontents*, Random House, 2020.

Milton Keynes UK
Ingram Content Group UK Ltd.
UKHW051603280324
440097UK00005B/21